THEORY AND COMPUTATION FOR SYNCHROTRON RADIATION SPECTROSCOPY

Related Titles from AIP Conference Proceedings

521 Synchrotron Radiation Instrumentation: Eleventh US National Conference
Edited by Piero Pianetta, John Arthur, and Sean Brennan, May 2000, 1-56396-941-6

505 Shock Compression of Condensed Matter—1999
Edited by M. D. Furnish, L. C. Chhabildas, and R. S. Hixson, April 2000,
2 vol. hard cover set, CD ROM included, 1-56396-923-8

489 Physics of Glasses: Structure and Dynamics
Edited by Philippe Jund and Rémi Jullien, October 1999, 1-56396-903-3

475 Applications of Accelerators in Research and Industry: Proceedings of the Fifteenth International Conference
Edited by J. L. Duggan and I. L. Morgan, June 1999, 2 vol. hard cover set,
1-56396-825-8

463 Photoacoustic and Photothermal Phenomena: Tenth International Conference
Edited by F. Scudieri and M. Bertolotti, March 1999, 1-56396-805-3

417 Synchrotron Radiation Instrumentation: Tenth US National Conference
Edited by E. Fontes, December 1997, 1-56396-742-1

To learn more about these titles, or the AIP Conference Proceedings Series, please visit the webpage **http://www.aip.org/catalog/aboutconf.html**

THEORY AND COMPUTATION FOR SYNCHROTRON RADIATION SPECTROSCOPY

Frascati, Italy   23–25 September 1999

EDITORS
M. Benfatto
C. R. Natoli
E. Pace
*Laboratori Nazionali di Frascati
dell'INFN, Italy*

Melville, New York
AIP CONFERENCE PROCEEDINGS ■ 514

Editors:

M. Benfatto
C. R. Natoli
E. Pace

INFN
Via E. Fermi 40
I-00044 Frascati
ITALY

E-mail: maurizio.benfatto@lnf.infn.it
calogero.natoli@lnf.infn.it
elisabetta.pace@lnf.infn.it

Authorization to photocopy items for internal or personal use, beyond the free copying permitted under the 1978 U.S. Copyright Law (see statement below), is granted by the American Institute of Physics for users registered with the Copyright Clearance Center (CCC) Transactional Reporting Service, provided that the base fee of $17.00 per copy is paid directly to CCC, 222 Rosewood Drive, Danvers, MA 01923. For those organizations that have been granted a photocopy license by CCC, a separate system of payment has been arranged. The fee code for users of the Transactional Reporting Service is: 1-56396-936-X/00/$17.00.

© 2000 American Institute of Physics

Individual readers of this volume and nonprofit libraries, acting for them, are permitted to make fair use of the material in it, such as copying an article for use in teaching or research. Permission is granted to quote from this volume in scientific work with the customary acknowledgment of the source. To reprint a figure, table, or other excerpt requires the consent of one of the original authors and notification to AIP. Republication or systematic or multiple reproduction of any material in this volume is permitted only under license from AIP. Address inquiries to Office of Rights and Permissions, Suite 1NO1, 2 Huntington Quadrangle, Melville, N.Y. 11747-4502; phone: 516-576-2268; fax: 516-576-2450; e-mail: rights@aip.org.

L.C. Catalog Card No. 00-101911
ISBN 1-56396-936-X
ISSN 0094-243X
Printed in the United States of America

CONTENTS

Preface vii

PART 1
PROBING THE ELECTRONIC PROPERTIES

Multiplet Effects in Resonant X-Ray Emission 3
    F. M. F. de Groot
Polarization Dependence in Resonant X-Ray Emission of TiO_2 13
    A. Kotani
Charge Ordering and Forbidden Reflections in Magnetite 20
    J. Garcia, G. Subias, M. G. Proietti, J. Blasco, H. Renevier,
    J. L. Hodeau, Y. Joly, and M. C. Sanchez
Orbital Ordering and Resonant Diffraction in Manganites.................. 30
    Y. Joly, M. Benfatto, and C. R. Natoli
Orbital Ordering and Metal-Insulator Transition in V_2O_3 45
    M. Cuozzo, Y. Joly, E. K. Hlil, and C. R. Natoli
Quantum Monte-Carlo Calculations and Possible Impact
on Angle Resolved Photoemission Spectroscopy 57
    W. Schattke
Photoemission Revealing Signature of Stripes and Orbital Modulation
in High T_c Superconductors 74
    N. L. Saini and A. Bianconi
The GW Approximation: Theory and Application to YH_3.................. 85
    T. Miyake, F. Aryasetiawan, H. Kino, and K. Terakura
Is a Hole a Single Particle?... 97
    C. Kim
Calculation and Interpretation of X-Ray Spectroscopies with Green's
Function Multiple Scattering Theory 105
    A. L. Ankudinov, A. Nesvishskii, and J. J. Rehr
Treatment of Non-Collinear Spin-Structures in Photo Emission
and X-Ray Absorption.. 110
    H. Ebert, J. Minár, V. Popescu, L. Sandratskii, and A. Mavromaras

PART 2
PROBING THE STRUCTURAL PROPERTIES

A New Recursive Approach to Photoelectron Diffraction Simulation 123
    F. J. Garcia de Abajo, M. A. Van Hove, and C. S. Fadley
Direct Methods for Surface X-Ray Diffraction 130
    D. K. Saldin, R. Harder, V. L. Shneerson, H. Vogler, and W. Moritz

On the Temperature Dependence of Multiple- and Single-Scattering Contributions in Magnetic EXAFS 140
    H. Wende, F. Wilhelm, P. Poulopoulos, K. Baberschke, J. W. Freeland, Y. U. Idzerda, A. Rogalev, D. L. Schlagel, T. A. Lograsso, and D. Arvanitis

EXAFS and Thermal Expansion 148
    G. Dalba, P. Fornasini, R. Grisenti, and F. Rocca

Dynamical Scattering of X-Ray by Real Binary Crystals and Problem of Point Defects 153
    L. I. Datsenko, V. P. Klad'ko, V. F. Machulin, S. Manninen, and I. V. Prokopopenko

Author Index 163

PREFACE

This volume contains papers presented at the Workshop on Theory and Computation for Synchrotron Radiation, which was held on 23-25 September, 1999 at the Laboratori nazionali di Frascati (Italy) as a sequel to those held in 1997 at Berkeley and in 1998 at Argonne.

The purpose of the Workshop was to dicuss the up-to-date theoretical understanding of Synchrotron Radiation Spectroscopies through a strong interaction between theorists and experimentalists and to highlight the advances in the fundamental aspects of the theory as well as the developments of efficient theoretical methods to calculate and simulate experiments. Moreover two sessions of the workshop were devoted to the establishment of a global world-wide Synchrotron Radiation Research Theory Network, whose scope and role is illustrated at the Network web site: http://www.cse.clrc.ac.uk/Activity/SRRTNet.

Past experience has shown that the sophistication of present day spectroscopies has reached a point at which a simple interpretation of the data is no more sufficient to extract all the wealth of information present therein. In some cases, like EXAFS analysis of absorption spectra or photoelectron diffraction in surface physics, the possibility of calculating a good theoretical signal to compare with experimental data has been of paramount importance to establish reliable tools for structural analysis. The theoretical success for these two spectroscopies is due to the fact that they are asymptotic theories in the energy parameter. In this case simple approximations are usually sufficient for a realistic description of the general system at hand, like the one electron approximation moving in an optical potential of the local density type and the muffin-tin approximation for its geometrical shape. Unfortunately, while structural information dominates the high energy spectral region, both structural and especially electronic information are present in the low energy region and it would be very important and extremely useful to have a reliable theory to help in the analysis of the data for the extraction of the relevant information. All the more that the most interesting and promising spectroscopies to unravel the electronic structure, like X-ray magnetic and natural dichroism, resonant elastic and inelastic scattering, both magnetic and non magnetic, give the most of electronic information near the excitation edge. The same is true, *mutatis mutandis*, for X-ray emission spectroscopies or angular resolved photoemission spectra from valence band in correlated electron systems.

In this energy region, complicated many-body processes intervene to screen the core (valence) hole and interfere with the direct excitation channel. Indeed, electron correlation effects in the final state and the coupling of the excited photoelectron with the photoinduced hole concur to complicate the description of the near edge region in many spectroscopies. Now, despite many efforts in the past to tackle these problems, they have not yet received a satisfactory solution. This is the reason why in many spectroscopies the near edge region is not adequately exploited to extract the rich information, both structural and especially electronic, present in it. The Frascati meeting was intended to give a contribution toward the solution of these longstanding problems in order to forge the tools for an efficient exploitation of the sophisticated spectroscopies using the electromagnetic radiation generated by third generation storage rings.

In organizing the workshop we had to choose from a rich and varied field of photon spectroscopies. In doing so, perhaps rather arbitrarily, we laid emphasis on the role of electronic correlations and on the theoretical analysis of the experimental results coming from elastic and inelastic resonant X-ray scattering, resonant X-ray emission, angle resolved photoemission spectroscopy, which have shed new light on the electronic structure of many interesting compounds. The papers dealing with these spectroscopies constitute the content of part I of this volume. However we have not discarded more traditional spectroscopies, like photoelectron diffraction, surface diffraction, magnetic EXAFS, EXAFS and thermal expansion and dynamical scattering of X-rays, when new advances have been made. Correspondently the papers dealing with these issues form the content of part II. In both cases we believe that this collection of stimulating papers represents the forefront of the current research effort in the field.

The organizers of the workshop gratefully aknowledge financial support by Istituto Nazionale di Fisica Nucleare (INFN), Istituto Nazionale di Fisica della Materia (INFM), Societa' Italiana di Luce di Sincrotrone (SILS), the Synchrotron Radiation Facilities of Trieste (Elettra), Grenoble (ESRF), and Berkeley (ALS) and the Italian computing companies CIE Telematica, Datamax, and SGI.

On behalf of all the participants we would like to express our gratitude to the secretarial and technical staff of the Laboratori Nazionali di Frascati, Mrs. S. Giromini, Mrs. R. Ianni, Mrs. M. Legramante, and Mrs. R. Morani for their competent and enthusiastic secretarial work and Mr. C. Federici and Mr. G. Bernardi for their precious technical assistance. Finally, heartfelt thanks go to our collegues of the Local Organizing Committee, Dr. A. Balerna, Dr. F. Boscherini, Dr. A. Marcelli, and Prof. S. Mobilio, who actively helped in organizing the workshop, and to the members of the Scientific Committee (M. Altarelli, K. Baberschke, P. Carra, A. Fujimori, A. Kotani, N. Harrison, J.J. Rehr, G. Sawatzky, M.A. Van Hove, R. Vedrinskii), whose wisdom guided us in the choice of the scientific subjects.

The Editors

M. Benfatto
C. R. Natoli
E. Pace

PART 1
PROBING THE ELECTRONIC PROPERTIES

Multiplet effects in Resonant X-ray Emission

Frank M.F. de Groot

*Department of Inorganic Chemistry and Catalysis, Utrecht University,
Sorbonnelaan 16, 3584 CA Utrecht, the Netherlands.*

Abstract. After a short discussion of all conventional core level spectroscopies within the single particle model, the effects of the coupling of the core and valence wave function on the x-ray emission spectral shapes is discussed. It will be shown that these so-called multiplet effects strongly affect all x-ray emission spectra taken around the metal 2p resonances. In case of 1s resonances, valence band x-ray emission is not affected, but the spectral shapes of 1s2p and 1s3p x-ray emission can only be sensibly described with the inclusion of multiplets. A special example is the resonant excitation into the pre-edge region, which gives rise to a quadrupole resonance.

CORE LEVEL SPECTROSCOPIES

All core level spectroscopies will be introduced using a simplified single particle model. This model is not expected to give a correct interpretation of the spectral shapes and intensities observed, but serves as a starting point on which more complete models build upon.

The Single Particle Approximations

In order to get a first idea of how the x-ray and electron spectra will look like we use a simplified single particle model. A series of approximations have been made:

1. **Density-of-States approximation:** The assumption that the single particle Density-of-States (DOS), as calculated for example by Local Spin Density (LSD) calculations or by real space Multiple Scattering codes, gives a good account of the electronic structure of the ground state.

2. **Spectroscopy approximation:** The assumption that the ground state DOS can in fact be used to describe spectroscopic transitions. This implies, for example, that it is assumed that in photoemission the valence hole does not cause a redistribution of the electronic states.

3. **Core hole approximation:** The assumption that the core hole created in x-ray absorption does not modify the DOS.

4. **Matrix Element approximation:** The assumption that the transition matrix elements are constant over the energy range analyzed.

Some of these omitted effects are relatively easy to include. For example, it is in general straightforward to include the matrix element effects. In electronic structure models based on real-space multiple scattering, it is also straightforward to include the core hole. The closely connected approximations (1) and (2) are more difficult to include and are related to electron-correlation effects that are not included in mean-field electronic structure methods.

The Core Hole Spin-Orbit Splitting

The coupling of the orbital and the spin-moment is given by the spin-orbit interaction, which is essentially a relativistic effect. The spin-orbit interaction is large for core holes and in general two peaks or structures will be visible in the spectrum, separated by the core hole spin-orbit splitting. The relative intensity of the $2p_{1/2}$ and $2p_{3/2}$ peak is 1:2 (given by the degeneracy of the states) and that of the $3d_{3/2}$ to $3d_{5/2}$ peak is 2:3. This rule is general for all core level spectroscopies. The rule breaks down *only* if another interaction is able to mix the $2p_{1/2}$ and $2p_{3/2}$ states. It will be discussed below that an important interaction is the overlap of the core hole wave function with the valence state wave functions, or in other words, the coupling of the core and valence moments. This coupling destroys the simple picture sketched above.

X-ray Absorption and X-ray Photoemission

It has been shown that the x-ray absorption (XAS) process can be described with the dipole approximation, under which the orbital quantum number must be modified by one while the spin quantum number is conserved. This implies that if a 1s-core electron is excited one observes the empty DOS of p-character. This p-projected unoccupied DOS is abbreviated as C_p. Similarly the excitation of a p core electron probes C_s plus C_d. The Fermi level is reached if the x-ray energy $\hbar\omega$ is exactly equal to the binding energy of the core hole (E_{1s}). The unoccupied DOS is reproduced by the spectrum with energies $C=\hbar\omega+E_{1s}$, with the 1s binding energy given with a negative number.

The same core hole excitation process can excite an electron out of a solid. Its kinetic energy (E_k) can be detected and gives directly the binding energy (E_c) of the core electron as: $E_c=E_k-\hbar\omega$, with the binding energy given as a negative number (and omitting the effects of the work function). The 1s XPS spectral shape thus consists of a single line. If a valence electron (V) is excited, exactly the same formula implies: $E_c = E_k - \hbar\omega$. An electron at the Fermi level is detected if the kinetic energy of the emitted electron exactly equals the x-ray energy, i.e. the occupied DOS can be determined by subtracting the x-ray energy from the kinetic energy: $V=E_k-\hbar\omega$, again using negative numbers for the occupied DOS. In principle the dipole selection rule applies again, but because the emitted electron can have any angular momentum in practice the dipole selection rule is dysfunctional and one observes the total occupied DOS (within the imposed approximation on matrix elements). Inverse Photoemission (IPES) is the inverse process of photoemission: an electron is directed towards the sample, it is absorbed in the unoccupied DOS, thereby emitting an x-ray.

X-ray Emission and Auger

The core hole decay can take place radiatively by a x-ray emission (XES) process or non-radiatively by an Auger process (AES). XES follows the same selection rules as x-ray absorption and a 1s-core hole can be filled by a core 2p electron or a valence p electron. If a 2p core electron fills the 1s core hole the x-ray energy gives the difference in binding energy of the two core states $E_{2p} - E_{1s} = \hbar\omega'$. If a valence electron fills the 1s core hole one maps the p-projected occupied DOS: $V = \hbar\omega' + E_{1s}$.

Instead of the dipole matrix element r, decay of a core hole can occur via the electrostatic two-electron integrals $<ab|1/r|cd>$, where a, b, c and d are electronic states. An example is the matrix element $<1s\ \varepsilon_d|1/r|2p2p>$. In this Auger process one 2p electron fills the 1s core hole and the other 2p electron is excited out of the solid as a free electron. In this 1s2p2p (or $KL_{2,3}L_{2,3}$) Auger process the final state contains two 2p-core holes. The kinetic energy of the emitted electron equals $E_k = E_{2p} + E_{2p} - E_{1s}$. If a 2p-core electron and a valence electron take part in the Auger process one can detect again the occupied DOS. Neglecting the Auger matrix elements (selection rules), we assume that the total occupied DOS is detected: $V = E_k - E_{2p} + E_{1s}$. It is also possible that two valence electrons take part in the Auger process. In this 1sVV Auger process one detects the self-convolution of the occupied DOS: $2V = E_k + E_{1s}$.

Resonant PES, AES and IPES

In the foregoing, core hole creation (XAS) has been described separately from core hole decay (XES and AES). However as soon as core hole creation takes place decay occurs. This implies that close to the XAS absorption resonance's, the decay processes can be different from off-resonance excitations. The only effect within the single particle model is that the excited electron can take part in the decay process. This creates additional decay channels not present in off resonant or normal AES and XES.

The best-known resonant spectroscopy is resonant photoemission (R-PES). Within the single particle model this can be described as the two-step process of x-ray absorption followed by Auger, i.e. $\Phi_0 \to^{[XAS]} \to \underline{1s}C \to^{[AES]} \to V$, where Φ_0 is the ground state and $\underline{1s}$ a 1s core hole. In the final state, a hole exists in the valence band and one measures the occupied DOS as in a normal photoemission process, but a difference is caused by the matrix element. The direct PES channel and the indirect XAS+AES channel have the same initial and final states, hence they interfere with each other.

There is another possible R-PES channel, i.e. $\Phi_0 \to^{[XAS]} \to \underline{1s}C \to^{[AES]} \to V+V+C$. In this case the C electron of the intermediate state does not participate in the Auger decay and this process is called the spectator channel. The process in which the C does participate is called the participator channel. The final state of the spectator channel has two valence holes (V) plus an extra electron in C. This can be viewed as a normal photoemission final state, plus a V to C excitation. This final state cannot be reached by normal PES within the single particle model. It plays a crucial part of all many body descriptions of core level spectroscopy, i.e. it is essentially the shake-up channel. Going to off-resonance conditions, the spectator channel disappears, as the C electron becomes a free electron.

Another resonant photoemission process is for example 2p3pV R-PES, in which first a 2p-core hole is created at resonance that subsequently decays by Auger to a 3p-core hole, i.e. $\Phi_0 \rightarrow^{[XAS]} \rightarrow 2pC \rightarrow^{[AES]} \rightarrow 3p, 3p+V+C$. The final state in the participator channel is again equal to normal 3p excitation, while the spectator channel adds a V to C excitation. Resonant Auger spectroscopy (R-AES) is the same as R-PES. The name R-AES will be reserved however for those processes that cannot be reached by direct photoemission. In practice this are all final states with two core holes present in the final states. An example is the 2p3p3p R-AES process, in which first a 2p-core hole is created at resonance that subsequently decays by Auger to two 3p-core holes, i.e. $\Phi_0 \rightarrow^{[XAS]} \rightarrow 2pC \rightarrow^{[AES]} \rightarrow 3p3pC$. Only spectator R-AES processes exist, because if the C electron participates one never creates a two-hole final state. If one moves away from resonance, the created C electron becomes a free electron, implying that the R-AES channels disappear.

The reverse process, resonant inverse photoemission (R-IPES), is also possible. If one excites with electrons that have a kinetic energy equal to the binding energy of a core state, an Auger process can occur in which both the core electron and the impinging free electron are transferred to an electron in the conduction band. In a second step, a conduction electron can decay by XES to the core hole. This could again be called a participator channel. In the spectator channel a valence electron decays to the core hole, i.e. $\Phi_0 \rightarrow^{[AES]} \rightarrow 2pCC \rightarrow^{[XES]} \rightarrow C, C+C+V$.

Resonant XES

It is also possible to study XES processes at resonance. For example one can excite a 2p core electron and detect valence band XES. In that case 'participator' R-XES (R-XES) equals the resonant elastic scattering path, i.e. $\Phi_0 \rightarrow^{[XAS]} \rightarrow 2pC \rightarrow^{[XES]} \rightarrow \Phi_0, V+C$. One observes essentially dd-excitations in a 2p R-XES experiment. Because of its similarity to Raman spectroscopy, R-XES is also called Resonant X-ray Raman Spectroscopy (R-XRS). Yet another name, often used to stress the relation to off-resonant inelastic scattering, is Resonant Inelastic X-ray Scattering (R-IXS).

Another R-XES channel is 2p3s R-XES, i.e. $\Phi_0 \rightarrow^{[XAS]} \rightarrow 2pC \rightarrow^{[XES]} \rightarrow 3sC$. The final state consists of a 3s-core hole and an electron in the d-projected unoccupied DOS. The energy difference between the incoming ($\hbar\omega$) and emitted ($\hbar\omega'$) x-rays, plus the (negative) 3s binding energy, equals the energy of the C state: $C = E_{3s}+\hbar\omega-\hbar\omega'$. In other words 2p3s R-XES is just a complicated detection technique of 2p XAS (within the single particle model).

Table 1 contains all conventional core level spectroscopies and the results obtained within the single particle approximation. The spectral shape is given by the conservation of energy between initial and final states. For example, 1s XAS is given by $C - E_{1s} = \hbar\omega$. Square brackets indicate an alternative experiment. Boldface implies that this energy is varied to measure the spectrum and it can be seen immediately that XAS and XES are x-ray spectroscopies, while PES, IPES and AES are electron spectroscopies. All resonant experiments do contain two variable energy scales. Note that within the single-particle approximation, XAS, IPES, R-AES and R-IPES all have the conduction band structure as there result. The main reasons that in experiment,

XAS and IPES have different spectral shapes are the differences in matrix elements and the core hole effect.

TABLE 1. Core Level Spectroscopies in the single particle model

Spec.	Core Holes	Spectral Shapes	DOS	
XAS	1s	$C - E_{1s} = \hbar\omega$	C	
XPS	2p [V]	$E_{2p} [V] = E_k - \hbar\omega$	-	[V]
IPES	-	$C = E_k - \hbar\omega$	C	
XES	1s2p [1sV]	$E_{2p}-E_{1s}$ [V-E_{1s}]$= \hbar\omega'$	-	[V]
AES	1s2p2p [1sVV]	$2E_{2p}-E_{1s}$ [2V-E_{1s}]$= E_k$	-	[2V]
R-PES	1sVV [1sVC]	V [2V-C] $= E_k - \hbar\omega$	V	[2V-C]
R-AES	2p3p3p	$2E_{3p}-C = E_k - \hbar\omega$	C	
R-XES	2p3s [2pV]	$C-E_{3s}$ [C-V]$= \hbar\omega - \hbar\omega'$	C	[C-V]
R-IPES	2pC [2pV]	C [2C-V]$= E_k - \hbar\omega$	C	[2C-V]

BEYOND THE SINGLE PARTICLE MODEL

The single particle interpretation gives a good overview of the potential use of the various spectroscopies. One of the important applications of resonant experiments is in fact the determination of the limitations of the single particle models, or in other words, the study of many body effects. The many body effects related to the ground state electronic structure will not be discussed in this paper. I will focus only on another set of important interactions not included in the single particle model and being crucial in all states containing a core hole. This is the overlap of the core wave function and the valence wave function. The effects of this interaction are known as *multiplet effects*. In the remainder of this paper, the consequences of multiplet effects on XAS and resonant XES will be discussed. If we limit ourselves to core levels of the 3d-metals, multiplet effects are important in all cases where a 2p, 3p or 3s hole is present. In addition, the 3d-valence band is susceptible to strong correlation effects, but these will not be discussed in this paper. We consider the following cases:

- 2p XAS: multiplet effects in final state[1,2,3]
- 2p3s R-XES: multiplet effects in intermediate states and interference effects[4,5]
- 2p3d R-XES: dd-excitations and spin-flip excitations[6,7,8]
- 1sV XES: a probe of the valence band DOS[9]
- 1s2p R-XES: multiplet effects in the intermediate and final states[10]

2p x-ray absorption

In case of 2p x-ray absorption, the single particle model assumes the transition from the ground state to 2pC. The overlap of the 2p and 3d wave function is very strong and gives rise to energy effects of the order of 10 eV[1,2]. The spectral shape is completely dominated by the multiplet effects and any relation to a single particle interpretation is lost. For example, the 2p3d overlap makes the MnO spectrum look closely like the $2p^5 3d^6$ final state multiplet, implying that the ground state of MnO is dominated by the $3d^5$ configuration[3]. The multiplet effects can be viewed as a kind of 'energy microscope' that enlarges small initial state energy effects to much larger final state energy effects. As such meV energy differences can be made visible with 100 meV resolution[3].

2p3s resonant x-ray emission

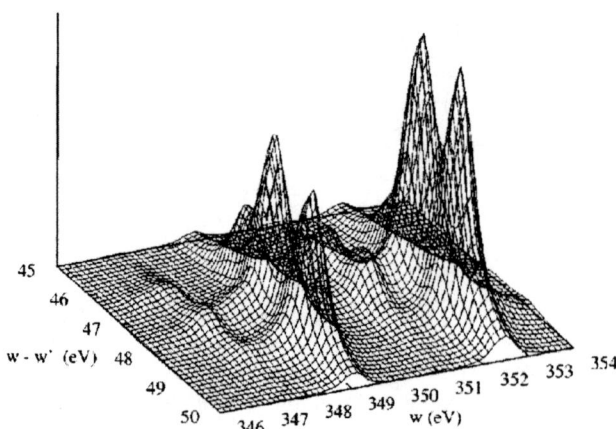

FIGURE 1. Intensity plot of the RIXS cross sections, calculated using Eq. 1, that is including interference effects. Excitation energy (ω) on the x axis; final-state energy (ω-ω') on the y axis. [reproduced from reference 5].

In case of 2p3s R-XES, the intermediate state is strongly affected by multiplet effects. In general a large number of intermediate states are present, which can give rise to complicated interference effects because the life time broadening is equivalent in size to the splitting between the levels. In the next section we discuss the complex case of Mn, but here we focus on the simplest case, being that of a system with an empty 3d band in the ground state, such as Ca^{2+} compounds as CaF_2. The ground state has a $3d^0$ configuration, implying a final state of 2p x-ray absorption with a $2p^53d^1$ configuration. The transitions can be essentially calculated by hand. It has been shown that this 2p x-ray absorption spectrum consists of seven final states, which are well separated in the experiment[1]. On each of these seven peaks one can study the resonant 2p3s x-ray emission decay. The experimental results[4] have been simulated with multiplet calculations, using the Kramers-Heisenberg formula:

$$I(\omega,\omega') \sim \sum_f \left| \sum_x \frac{\langle \Phi_f | \hat{e}' \cdot r | \Phi_x \rangle \langle \Phi_x | \hat{e} \cdot r | \Phi_i \rangle}{\omega - E_x - i\Gamma_x} \right|^2 \delta_{E_f + \hbar\omega' - E_i - \hbar\omega} \qquad (1)$$

with ω being the incoming x-rays, ω' the emitted x-rays, Φ_i the initial state, Φ_x the intermediate state and Φ_f the final states. (See ref.[5] for details). The resonance part of the experiments can be nicely simulated with these calculations. The simulations can be represented as a two-dimensional plot of ω and ω' as given in Figure 1. In practice one uses the energy of the final states, being equal to the difference between ω and ω'. The final state energy gives the binding energy of the $3s^13d^1$ final states. The only multiplet effect active in this $3s^1 3d^1$ configuration is the 3s3d exchange interaction. The only other energy effect (within the model used) is the cubic crystal field, and four peaks are found split by these two interactions.

2p3d resonant x-ray emission

The first 2p3d R-XES study has been carried out by Butorin and coworkers[6]. They studied the 2p3d decay at the 2p x-ray absorption resonances of MnO. The experiments show a series of peaks at energy losses ($\omega-\omega'$) equal to the well known dd-excitations. The general trends of the data could already be explained from an atomic multiplet model[6]. The details of the dd-transitions and the charge transfer peaks can be explained by using the more complete charge transfer multiplet model[3,7]. By using this model, the ff-transitions in rare earth 3d4f R-XES spectra have been made visible. Details on these experiments can be found in a recent review article on RIXS[8].

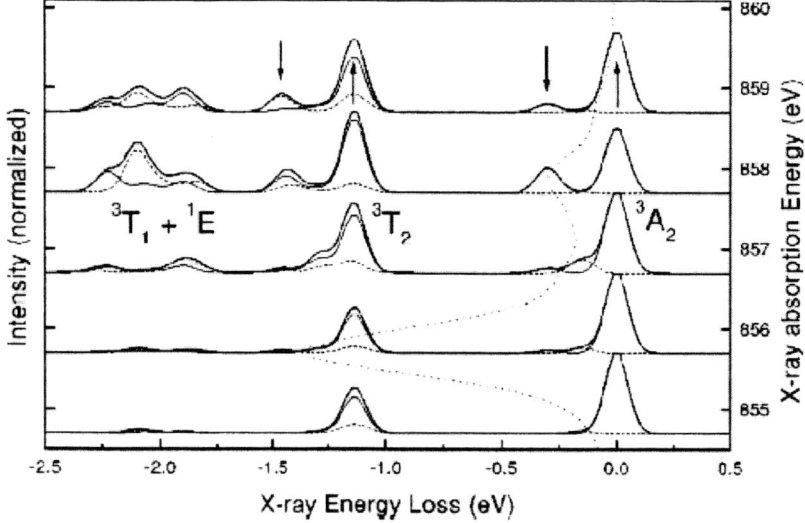

FIGURE 2. 2p3d RIXS at the 2p edge of a Ni^{2+} system in octahedral symmetry, using the crystal-field parameter of NiO. Indicated are F_{xx} scattering (thin solid), F_{zx} scattering (dashed), and the total scattering (thick solid). The x-ray-absorption spectrum is given with dots. The symmetries of the states are given at the middle spectrum. The M_S sub-states are indicated at the top for spin up and spin down (figure reproduced from Reference 7).

In this paper, I would like to focus on yet another phenomenon of 2p3d R-XES, the possibility to observe spin-flip transitions. The usage of the term spin-flip is not consistent in articles on R-XES/R-XRS/RIXS, so I will try to clarify the different usage's here. Sometimes part the dd-transitions are called spin-flip transitions because the total spin quantum number S of the configurations is changed from, for example, 5/2 to 3/2 as is the case for the MnO experiments. Instead of a change in S, also a change in M_S is called spin-flip. I will consider a change in M_S as a real spin-flip transition, the change in S being called a multiplet excitation, as it are the multiplet effects that cause the energies involved. Multiplet excitations are of the order of a few eV, whereas spin-flip excitations are of the order of 100 meV, being closely related to the superexchange energy[7,9]. With the improvement in the resolution of soft x-ray R-XES experiments, these 0.1 eV transitions can be made visible in experiment[9].

1sV x-ray emission

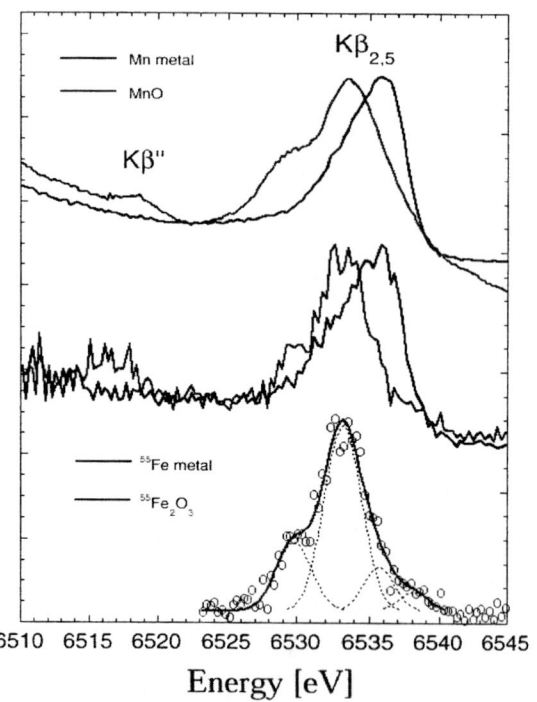

FIGURE 3. Valence band 1sV x-ray emission spectra. Top to bottom: (a) X-ray excited spectra for MnO (red) and Mn metal (black); (b) K-capture spectra for 55Fe$_2$O$_3$ (red) and Fe metal (black); (c) a fit of the background-subtracted valence band region.

In case of 1s x-ray absorption, the main edge is due to 1s to 4p transitions. Its spectral shape is hardly affected by multiplet effects. The situation is different for the pre-edges that are discussed in the next section. Valence band emission to the 1s core hole, on the main edge as well as off-resonance, leaves the final state with a valence hole and an essentially free electron that can be neglected in the description of the 1sV x-ray emission spectral shape. Thus, 1sV x-ray emission is not affected by multiplets and describes the valence band DOS, similar to valence band XPS. A difference is caused by the matrix elements that in case of 1sV x-ray emission allow transitions only from the p-projected DOS. In case of 3d-metal oxides, this implies the metal p-states in the oxygen 2p-valence band. In addition at threshold, weak quadrupole transitions from the occupied 3d-states can be expected, in a way similar to the pre-edge structures in 1s XAS, i.e. including the admixture effects of 4p and 3d character.

Figure 3 shows the valence band region (Kβ$_{2,5}$ in spectroscopic notation). The bottom spectrum shows the oxygen 2p valence band at 6530 and 6534 eV. The intensity around 6538 eV is expected to be due to the 3d valence band. The Kβ" peak at 6515 eV is due to manganese 4p-character in the oxygen 2s states. This peak is denoted as a cross-over peak as it can be viewed as a x-ray emission channel from oxygen to manganese. The top spectra excited with x-rays several tens of eV above the resonance. The middle spectrum shows x-ray emission spectra, where the core hole has been created by the K capture process. In K capture ^{55}Fe decays to ^{54}Mn by reaction of a 1s core electron with a proton to a neutron and neutrino. This creates a Mn atom with a 1s core hole, that subsequently decays similar to an x-ray excited Mn atom. For more details, including a discussion of the differences in x-ray emission spectral shapes between K capture and x-ray excited core holes, is referred to ref.[10].

1s2p resonant x-ray emission

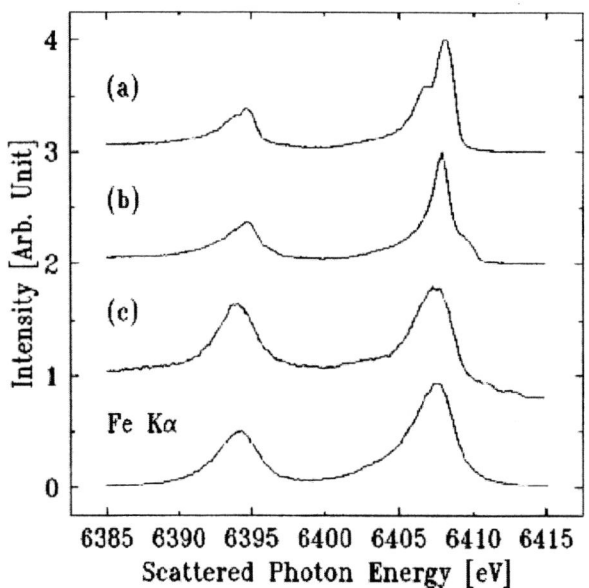

Figure 4: RIXS spectra as a function of the scattered photon energy for three different excitation energies together with the off resonance 1s2p x-ray emission spectrum. The excitation energies are 7113.4 (a), 7114.7 (b) and 7117.7 eV (c) for the RIXS spectra (reproduced from Ref. [11]).

As a last example, I would like to discuss the 1s2p x-ray emission spectra, taken at an excitation energy set exactly at the two peaks of the pre-edge structure of the 1s XAS spectrum of Fe_2O_3[11]. This pre-edge structure consists of two peaks at 7113.4 eV and at 7114.7 eV, split by the crystal field splitting of 1.3 eV. Both pre-edge peaks have an intensity of a few percent of the main edge and are related to 1s3d quadrupole transitions. This quadrupole excitation is followed by a dipolar 1s2p decay. The overall process can be accurately described with the multiplet model for the process

$3d^5 \rightarrow 1s^1 3d^6 \rightarrow 2p^5 3d^6$. Note that the final state has exactly the same configuration as would be the case for direct 2p x-ray absorption. Detailed analysis learns that this analogy is indeed correct and that the final state of this process are identical to that of 2p XAS, the only difference being different transition matrix elements[11].

CONCLUDING REMARKS

The examples given indicate that the x-ray emission processes that concern a 2p, 3s or 3p core level in the intermediate or final state are in all cases strongly affected by multiplet effects. In fact this concerns *all* resonant x-ray emission processes, except off-resonance 1sV XES. Multiplet analysis does yield an accurate simulation of all spectral shapes, and additionally provides valuable insights into the electronic structure of the materials studied (not discussed in this paper).

A very important usage of these R-XES processes is the measurement of XAS spectra by putting the detector at a particular final state. It has been shown[10] that a wide range of selective XAS spectra can be measured. This includes selectivity's for the valence, the spin-state and the type of neighbour[10,12].

As a last remark, I would like to mention that two recent review paper have appeared on soft x-ray resonant XES[8,13]. A paper by Butorin focuses on the multiplet analysis of transition metal and rare earth spectra[8]. A very extensive paper by Gel'mukhanov and Agren[13] can be considered 'orthogonal' to the present contribution as it concerns essentially all spectra for which multiplet effects are *not* important.

REFERENCES

1. de Groot, F.M.F, Fuggle, J.C, Thole, B.T. and Sawatzky, G.A., Phys. Rev. B. 41, 928 (1990)
2. de Groot, F.M.F, Fuggle, J.C, Thole, B.T. and Sawatzky, G.A., Phys. Rev. B. 42, 5458 (1990)
3. de Groot, F.M.F. J. Elec. Spec, 67, 529 (1994)
4. Rubensson, J.E., Eisebitt, S., Nicodemus, M., Böske, T, and Eberhardt, W., Phys. Rev. B 49, 1507 (1994)
5. de Groot, F.M.F., Phys. Rev. B. 53, 7099.
6. Butorin, S.M., Guo, J., Magnuson, M. Kuiper, P and Nordgren, J., Phys. Rev. B. 54, 4405 (1996).
7. de Groot F.M.F, Kuiper, P and Sawatzky, G.A. Phys. Rev. B. 57, 14584 (1998).
8. Butorin, S.M., J. Elec. Spec. (in press).
9. Kuiper, P., Guo, J., Såthe, C., Duda, L.C., Nordgren, J., Pothuizen, J.J.M., de Groot, F.M.F. and Sawatzky, G.A., Phys. Rev. Lett. 80, 5204 (1998).
10. Bergmann U, Glatzel P, Degroot F.M.F. and Cramer S.P., J. Am. Chem. Soc. 121, 4926 (1999).
11. Caliebe, W.A., Kao, C.C., Hastings, J.B., Taguchi, M., Kotani, A. Uozumi, T., and de Groot, F.M.F., Phys. Rev. B. 58, 13452 (1998).
12. de Groot, F.M.F., Topic in Catalysis (in press).
13. Gel'mukhanov, F. and Agren, H., Physics Reports 312, 87-330 (1999).

Polarization Dependence in Resonant X-Ray Emission of TiO$_2$

Akio Kotani

Institute for Solid State Physics, University of Tokyo,
Roppongi, Minato-ku, Tokyo 106-8666, Japan

Abstract. Theoretical investigation is made for the polarization dependence of resonant X-ray emission spectroscopy (RXES) of TiO$_2$. In Ti $2p \to 3d \to 2p$ RXES, a dramatic resonant enhancement of a 13 eV inelastic peak is predicted with TiO$_6$ cluster model, when the incident photon energy is tuned to a satellite of the Ti $2p$ XAS. A strong polarization dependence of this RXES is shown in the calculation, and confirmed by recent experimental observations. These results suggest strongly that the satellite of the Ti $2p$ XAS originates from the excitation to the anti-bonding state between the Ti $3d$ and O $2p$ states due to the covalency hybridization, i.e., it is the charge transfer satellite.

INTRODUCTION

Resonant X-ray emission spectroscopy (RXES) is one of the most powerful tools in solid state spectroscopy with synchrotron radiation. In this paper we report new theoretical results on the polarization dependence of RXES, which gives quite important information on the mechanism of structures in X-ray absorption spectroscopy (XAS). In Ti $2p$ XAS of TiO$_2$ (rutile), a weak satellite is observed about 13 eV from the main peak [1], but its mechanism seems not to be established well. Okada and Kotani [2] proposed a charge transfer (CT) mechanism for this satellite, but other interpretations had also been proposed: van der Laan [1] proposed that the satellite originates from the simultaneous excitation of an O $2p$ - O $3s$ valence exciton (denoted by "polaronic" satellite), while a molecular-orbital diagram suggested that it would be asigned as an excitation of the anti-bonding orbital of Ti $4sp$ and O $2p$ character [3,4].

In the upper panel of Figure 1, the calculated result of XAS with a TiO$_6$ cluster model is shown, which is almost the same as that by Okada and Kotani [2]. In this calculation the satellite (shown with the arrow) is caused by the CT excitation from O $2p$ to Ti $3d$ states. If this mechanism is correct, it is expected that a strong enhancement of the CT excitation in RXES should be observed, when the incident photon energy is tuned to the CT satellite of XAS. In previously published RXES experiments, however, no enhancement was observed for the CT excitation [5,6]. In the following we will make it clear that the effect of the polarization dependence of RXES is essential in discussing the resonant enhancement of the spectra [7]. This is actually confirmed by recent experimental observations [8].

In Sec. 2 we calculate the Ti $2p \rightarrow 3d \rightarrow 2p$ RXES spectra of TiO_2, taking into account the effect of different polarization directions of the incident X-ray. We show that the polarization-dependence in the RXES is very important. In Sec. 3 we compare the calculated and experimental results, and give some discussions.

CALCULATIONS

We consider the geometrical configuration of the RXES observation as shown in Figure 2. The angle between the incident and emitted photon directions is fixed at 90°, and the incident photon polarization is in either of the y or z direction, which is denoted by the "polarized" or "depolarized" configuration, respectively (Figure 2 corresponds to the depolarized configuration). The polarization of the

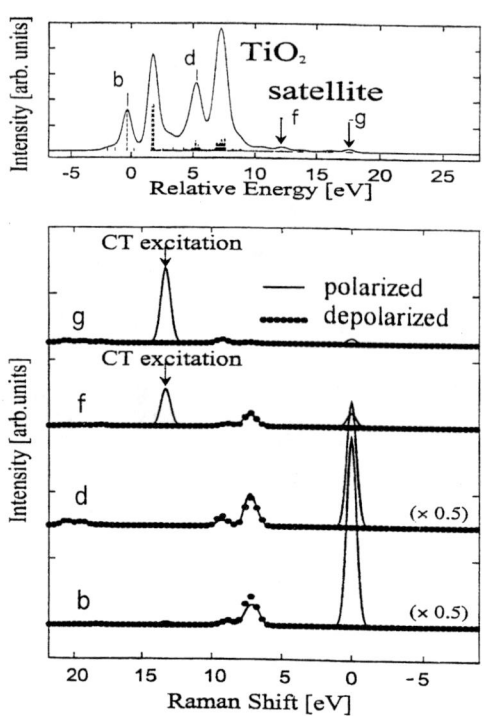

FIGURE 1. Calculated results of Ti $2p$ XAS and $2p \rightarrow 3d \rightarrow 2p$ RXES of TiO_2 with TiO_6 cluster model.

emitted photon is not detected. In these two configurations, the spectra of RXES are expressed as

$$F(\Omega, \omega) = \sum_{T_2=x,y} \sum_j \left| \sum_i \frac{<j|T_2|i><i|T_1|g>}{E_g + \Omega - E_i - i\Gamma} \right|^2 \delta(E_g + \Omega - E_j - \omega), \quad (1)$$

where Ω and ω are, respectively, the incident and emitted photon energies, $|g>$, $|i>$, and $|j>$ are initial, intermediate, and final states of the material system, respectively, E_g, E_i, and E_j are their energies, and Γ represents the spectral broadening due to the core-hole lifetime in the intermediate state. In the polarized and depolarized configutrations, the dipole transition operator T_1 is proportional to y and z, respectively.

The calculation of $F(\Omega, \omega)$ is made with the TiO_6 cluster model with O_h symmetry [7]. The model and the parameter values are almost the same as those in Ref. [2]. The results are shown in Figure 1 (lower panel), where the incident photon energy is taken at the positions **b**, **d**, **f** and **g** in the XAS spectrum (upper panel), and the RXES spectra are plotted as a function of $\Omega - \omega$ (denoted by Raman shift), where the elastic (Rayleigh) scattering line is at the origin. The spectral structure is divided into three categories: (1) elastic line at 0 eV, (2) inelastic spectra at 7 and 9 eV, and (3) inelastic line at 13 eV. The origin of these spectra can be explained by the energy level scheme shown in Figure 3. TiO_2 is nominally in the $3d^0$ state, but actually the $3d^0$ configulation is strongly mixed with a charge transferred $3d^1\underline{L}$ configulation by the covalency hybridization, where $\underline{L}$ denotes a hole in the ligand state (O $2p$ molecular orbit). The ground state is the bonding state between the $3d^0$ and $3d^1\underline{L}$ configulations, and the anti-bonding state is located about 13 eV above the ground state. Both bonding and anti-bonding states are specified by an irreducible representation A_{1g} of the O_h symmetry group. In addition to these states, there are non-bonding $3d^1\underline{L}$ states with $T_{1g}, T_{2g}, E_g \cdots$ symmetries about 7 $\sim$ 9 eV above the ground state. When a Ti $2p$ electron is excited to the $3d$ state by the incident photon, we have $2p^53d^1$ and $2p^53d^2\underline{L}$ configurations which are mixed

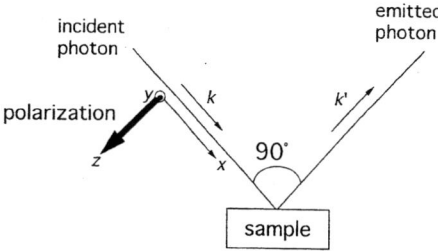

FIGURE 2. Geometry of the depolarized configuration. In the polarized configuration the polarization of the incident X-ray is taken in the y direction.

strongly by the covalency hybridization. The main peak of the Ti 2p XAS corresponds to the bonding state between the $2p^53d^1$ and $2p^53d^2\underline{L}$ configurations, while the satellite corresponds to the anti-bonding state between them. The intensity of the satellite is very weak because of the phase cancellation between the wave functions of the ground and photo-excited states [2]. Also, the X-ray absorption is almost forbidden to the non-bonbonding $2p^53d^2\underline{L}$ states. In Figure 3, we disregard the effects of the spin-orbit splitting of the 2p states and the crystal field splitting of the 3d states, for simplicity. If we take into account these effects, the main peak (and also the satellite) splits into four peaks.

The resonantly excited intermediate states, which correspond to the main peak and the satellite of the XAS, decay radiatively to each of the final states of RXES, i.e., the bonding, non-bonding, and anti-bonding states. The categories (1), (2), and (3) of RXES spectra correspond to the bonding, non-bonding, and anti-bonding final states, respectively. The calculated spectrum (2) has two peaks, which correspond to the crystal field splitting of the non-bonding $3d^1\underline{L}$ configuration $(3d^1(t_{2g})\underline{L}$

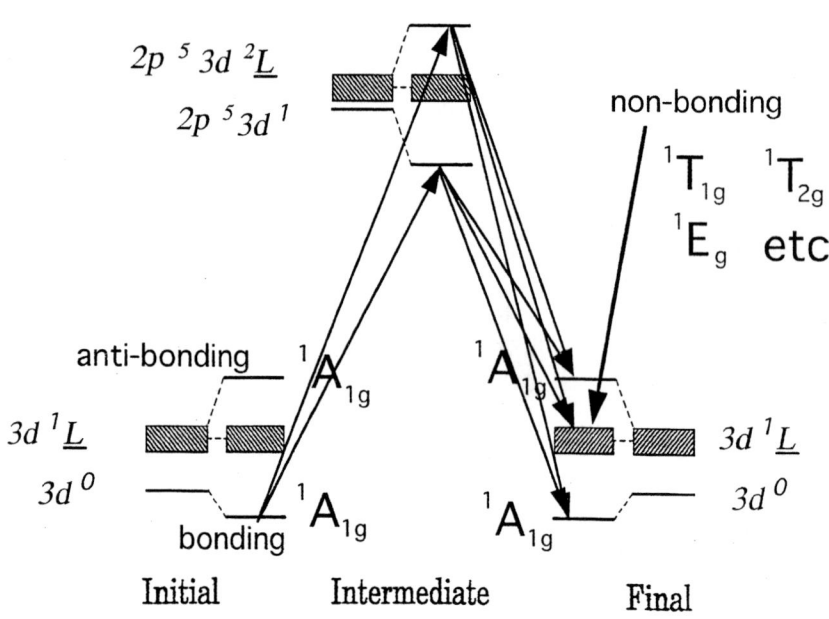

FIGURE 3. Schematic representation of the Ti $2p \rightarrow 3d \rightarrow 2p$ RXES of TiO_2 with the TiO_6 cluster model.

and $3d^1(e_g)\underline{L}$). The spectrum (3) occurs for the incident photon energy tuned to the satellite of the XAS spectrum. This is because the XAS satellite corresponds to the anti-bonding intermediate state of RXES, so that it enhances dramatically the intensity of the anti-bonding final state (3). The spectra (1) and (3) are allowed only for the polarized configuration, while the spectrum (2) is allowed for both polarized and depolarized configurations. The reason for this will be discussed in the next section.

COMPARISON WITH EXPERIMENTS AND DISCUSSION

Very recently, experimental measurements [8] of RXES for TiO_2 have been made with the polarized and depolarized configurations. The result is shown in Figure 4. The three categories of RXES spectra (1) $\sim$ (3) are clearly seen, in addition to the spectra indicated by vertical bars, which are absent in our calculated results

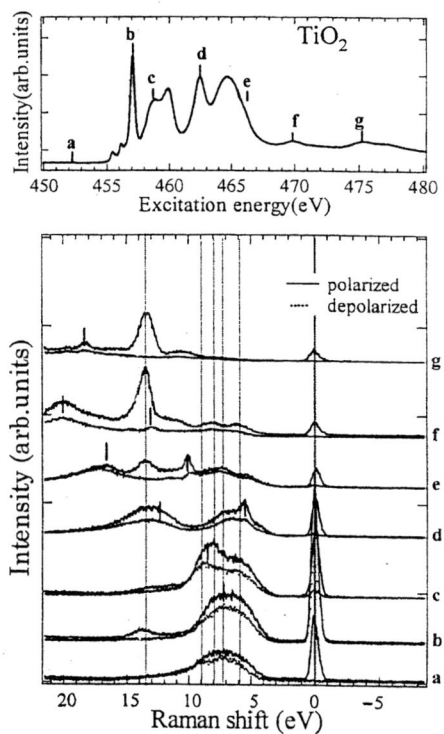

FIGURE 4. Experimental results of Ti $2p$ XAS and $2p \to 3d \to 2p$ RXES of TiO_2.

(Figure 3). The elastic scattering peak at 0 eV (category 1) and the inelastic one at 13 eV (category 3) are allowed only for the polarized configuration, and the intensity of the 13 eV peak is dramatically enhanced when the incident photon energy is tuned to the satellite of the XAS spectrum. Near the middle of the elastic (0 eV) and inelastic (13 eV) scattering peaks, there are inelastic scattering spectra (category 2) which are allowed both for the polarized and depolarized configurations. These results are in good agreement with the calculated ones, and strongly support the CT mechanism of the XAS satellite. It is to be remarked that the previous RXES measurements for TiO_2 were made only in the depolarized configuration, so that the resonant enhancement of the 13 eV inelastic peak could not be observed. The spectral width of (2) is much larger than that of the calculated result, and this broadening comes mainly from the energy band width of the O $2p$ states, which is desregarded in the cluster model.

Let us consider the mechanism of the polarization dependence of the RXES spectra. It is easily shown from Eq.(1) and group theoretical consideration that the final states with A_{1g}, T_{1g}, T_{2g}, and E_g irreducible representations are allowed for the polarized configuration, whereas those with only T_{1g} and T_{2g} irreducible representations are allowed for the depolarized configuration. Therefore, the elastic peak (bonding state) and the 13 eV inelastic peak (anti-bonding state) are allowed for the polarized configuration, but they are forbidden for the depolarized configuration. The nonbonding states are allowed both for the polarized and depolarized configurations.

It is to be mentioned that some difference between the calculated and experimental results is seen for the Ti $2p$ XAS: The second peak of the main structure is split into two in the experiment, but only one peak is seen in the calculation. This discrepancy is due to our approximation that the local symmetry around Ti is treated as O_h, but actually it is D_{2h}. The lower symmetry calculation reproduces this splitting correctly [9].

In the rest of this section, we discuss the origin of the spectra indicated by the vertical bars in Figure 4. The energy position of these bars changes almost proportionally to the change of the incident photon energy, so that the emitted photon energy is almost independent of the incident photon energy, similarly to the so-called normal X-ray emission spectroscopy (NXES) which is usually observed for the incident photon energy well above the XAS threshold. In the experimental result in Figure 4, however, these spectra are observed near the XAS threshold, so that we call them "NXES-like spectra". Recently Idé and Kotani calculated RXES with a one-dimensional d-p model (a simplified version of the nominally $3d^0$ system) with multi-transition-metal sites [10]. They showed that NXES-like spectra are absent in the cluster with a single transition metal site, but for larger clusters NXES-like spectra can occur near the XAS threshold because of the existence of spatially extended XAS final states (RXES intermediate states) due to the effect of of multi-transition-metal sites. Therefore, in order to reproduce the NXES-like spectra of TiO_2, it would be necessary to extend the cluster size to that larger than the TiO_6 cluster.

ACKNOWLEDGMENTS

The author would like to thank Mr. M. Matsubara, Dr. T. Uozumi, Professor K. Okada, Dr. H. Ogasawara, Mr. Y. Harada, and Professor S. Shin for collaborations and discussions. This work is partly supported by a Grand-in-Aid for Scientific Research from the Ministry of Education, Science, Sports and Culture in Japan. The computation in this work was done using the facilities of the Super-Computer Center, Institute for Solid State Physics, University of Tokyo.

REFERENCES

1. van der Laan, G., *Phys. Rev. B*, **41**, 12366 (1990).
2. Okada, K., and Kotani, A., *J. Electron Spectrosc. Relat. Phenom.* **62**, 131 (1993).
3. Fisher, D. W., *Phys. Rev. B* **5**, 4219 (1972).
4. Sen, S. K., Riga, J., and Verbist, J., *Chem. Phys. Lett.* **39**, 560 (1976).
5. Tezuka, Y., Shin, S., Agui, A., Fujisawa, M., and Ishii, T., *J. Phys. Soc. Jpn.* **65**, 312 (1996).
6. Jiménez-Mier, J., van Ek, J., Ederer, D. L., Callcott, T. A., Jia, J., Carlisle, J., Terminello, L., Asfaw, A., and Perera, R. C., *Phys. Rev.* **59**, 2649 (1999).
7. Matsubara, M., Uozumi, T., Kotani, A., Harada, Y., and Shin, S., unpublished.
8. Harada, Y., Kinugasa, T., Eguchi, R., Matsubara, M., Kotani, A., Shin, S., Watanabe, M., and Yagishita, A., unpublished.
9. de Groot, F. M. F., Fuggle, J. C., Thole, B. T., and Sawatzky, G. A., Phys. Rev. B **41**, 928 (1990).
10. Idé, T., and Kotani, A., J. Phys. Soc. Jpn. **67**, 3621 (1998).

Charge ordering and forbidden reflections in magnetite

J. Garcia*, G. Subias*, M. G. Proietti*, J. Blasco*, H. Renevier[†],
J.L. Hodeau[†], Y. Joly[†] and M.C. Sanchez*.

*Instituto de Ciencia de Materiales de Aragon. CSIC-Universidad de Zaragoza, plaza San
Franciso s/n, 50009 Zaragoza, Spain[1]
[†]Laboratoire de Cristallographie, CNRS, B.P. 166, F-38042 Grenoble Cedex 9, France

Abstract. Resonant scattering experiments at the iron K edge of the (002) and (006) forbidden reflections in magnetite are shown and discussed. The energy dependence of the intensity of these reflections, their dependence on the azimuthal angle and the polarization analysis are given. Experiments have been performed at room temperature and at low temperatures, below the Verwey transition. An identical behaviour has been observed at high and at low temperatures, showing that the same kind of local anisotropy is present in the two phases. This experiment demonstrates the absence of either charge fluctuation between octahedral atoms in the metallic phase, or charge ordering in the insulating phase.

INTRODUCTION

In the last years, transition metal oxides have been object of a tremendous interest due to the large variety of properties that they exhibit. As for example, high T_c superconductivity in copper oxides, colossal magnetoresistance in manganese oxides, metallic state and metal insulator phase transitions [1]. In particular, a lot of interesting properties have appeared in the mixed valence oxides, i.e. oxides where the formal valence state of the transition metal atoms is a non-integer number. In spite of the general classification given by Robin and Day [2] about the mixed valence compounds, establishing no charge localization for ions in equivalent lattice sites, it is generally assumed atomic electron localization in mixed valence oxides. This localization can be temporal or spatial, giving rise to several models of electrical conduction, based on charge fluctuation among different crystallographic sites and on the development of periodically ordered localized states, the so-called charge ordering phase, at low temperatures. There are very few cases in which charge localization seems to be directly observed [3]. Generally, the absence of metallic conduction in mixed valence oxides has been considered as a proof of

---

[1]) Supported by Spanish CICYT project n MAT 99-0847 and the LEA-MANES project.

atomic charge localization.

A direct method to determine the atom valence state is X-ray Absorption Spectroscopy, due to the dependence of the absorption energy edge on the valence state of the atom, the so-called chemical shift. This property has been mainly used to study the localization of f states in lanthanide compounds for both, spatial charge localization and fluctuating systems [4]. The atomic anomalous scattering factor is directly related to the X-ray absorption spectroscopy. Several experiments have been performed to determine the valence state of an atom by using anomalous diffraction technique [5]. It is particularly interesting the case of reflections forbidden by symmetry, where the structure factor is given by the difference of atomic scattering factors of crystallographic equivalent sites. In this case, forbidden reflections can be observed due to the anysotropy of the anomalous scattering factor (ATS reflections), which depends on the orientation of the X-ray polarization vector respect to the anisotropic axis and on the momentum transfer [6]. Therefore, if atoms are anysotropic with different orientation of the anysotropic axis in the unit cell, the structure factor for these reflections will be different from zero. This kind of reflections, that are considered as a probe of orbital ordering, also shows azimuthal periodicity, i.e. their intensity depends on the azimuthal angle of the reflection plane.

Another possibility, to get non-zero intensity forbidden reflections is the appearance of charge ordering. In this case, the anomalous scattering factors of the two valence states close to the absorption edge are different and consequently, we observe scattered intensity at the absorption edge energy due to the chemical shift between these two valence states. Therefore, the study of forbidden reflections can give us direct information on the electronic state of atoms and can solve the problem of electronic localization. One of the mixed valence compound where charge localization has been proposed is magnetite. The observation and analysis of (0 0 2+4n) forbidden reflections will give us direct information of the electronic state of the iron atoms.

Magnetite belongs to the family of oxides that crystallize with the spinel cubic structure, $MgAl_2O_4$, space group $Fd\bar{3}m$. The cubic unit cell, illustrated in fig. 1, contains eight formula units, $A(B_2)O_4$, where the A sites are tetrahedral and the B sites are octahedral. The formula unit can be written as $Fe^{3+}(Fe^{2+}, Fe^{3+})O_4$, considering an ionic crystal. Here, the bracket indicates atoms located in the B (16d) site of the spinel lattice, while the rest of cations are in the A (8a) site. According to this formula, 2+ and 3+ ions coexist in the same B site. Fe_3O_4 is metallic at room temperature and at about 125 K develops a transition to a semiconducting state. This phase transition, known as Verwey transition, was postulated to be an order-disorder transition of the octahedral Fe^{3+} and Fe^{2+} ions [7]. Within this model, the electrical conductivity in the high temperature phase is explained in terms of a dynamical transformation of Fe^{3+} to Fe^{2+}, i.e one electron jumping between different octahedral B sites. The insulator behaviour, below the Verwey transition, is due to the localization of the movile electron on the octahedral B sites, giving rise to a charge ordered lattice. In spite of several experimental evidences againts this charge ordering model and the lack of consensus about the exact charge ordering arrangement at low temperature, a general opinion is that the conductivity in the high temperature phase is due to charge fluctuation between the octahedral iron sites. This charge localizes, showing long range order, on the octahedral atoms in the low temperature phase.

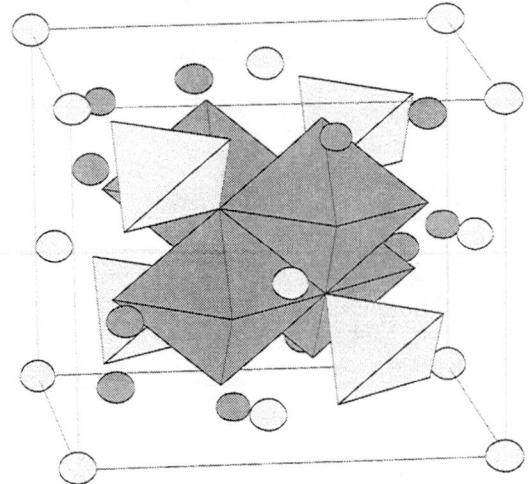

FIGURE 1. Crystallographic unit cell of magnetite. Only Fe atoms are shown. Clair balls are Fe atoms in the A site and dark balls in the B site. Both kind of polyedra are shown.

With the aim to find out what is the real charge ordering configuration in magnetite at low temperatures and the driving force for this charge ordering, we have carried out X-ray Resonant Scattering experiments at the iron K-edge at room and at low temperatures. We report here the observation of the (002) and (006) forbidden reflections in magnetite [8]. The analysis of these reflections demonstrates the absence of both, $Fe^{3+} - Fe^{2+}$ fluctuation at room temperature, and charge ordering at low temperatures.

EXPERIMENTAL DETAILS

A single crystal of magnetite was grown by the floating zone method using halogen lamps. The crystal was oriented and polished to obtain a flat clean (001) surface. The X-ray experiments were performed at the beamline D2AM at the European Synchrotron Radiation Facility (ESRF) in Grenoble. The low temperature measurements were performed with a closed cycle cryostat. The energy resolution of the incident beam (σ-polarized) was about 1 eV. σ-incident geometry was used in the experiment and the σ-scattered polarization analysis was performed using a MgO (222) crystal analyser.

RESULTS

Room temperature data

The (002) and (006) forbidden reflections have been measured at room temperature. The energy dependence at $\varphi=45$, where φ is the azimuthal angle, (incident polarization vector parallel to the [110] direction) is shown in fig.2.

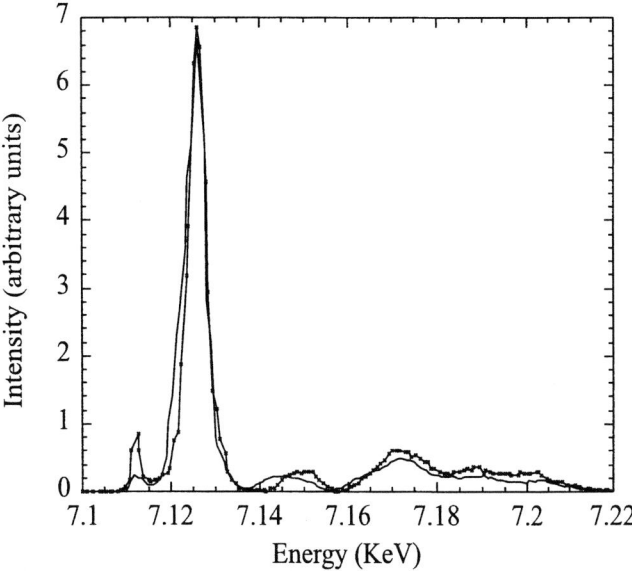

FIGURE 2. Comparison of the intensity versus photon energy of the (002) (solid line) and (006)(dots) forbidden reflections at room temperature. Spectra are corrected for self absorption.

No intensity has been observed below the iron K-edge energy, as expected for a forbidden Thompson reflection. Intensity is observed only above the absorption edge, so these reflections are only due to the anomalous part of the iron scattering factor. The energy dependence of the intensity does not change when varying φ, so the intensity can be factorized as $I = f(E)g(\theta, \varphi)$, where θ is the Bragg angle. Three main features can be distinguished: i) a resonance at the energy value of the prepeak in the fluorescence spectrum, ii) a main resonance at the iron absorption K-edge and iii) an oscillatory behaviour of the intensity at energies above the absorption edge. The main resonance and the extended part are very close to each other when comparing the (002) and (006) reflections, whereas the prepeak resonance is more intense for the (006) reflection. The intensity dependence on the azimuthal angle φ is shown in fig. 3. It is observed that the intensity shows a $\pi/2$ periodicity on φ and also depends on the Bragg angle.

The anomalous atomic scattering factor due to the resonance of a ground state **a**

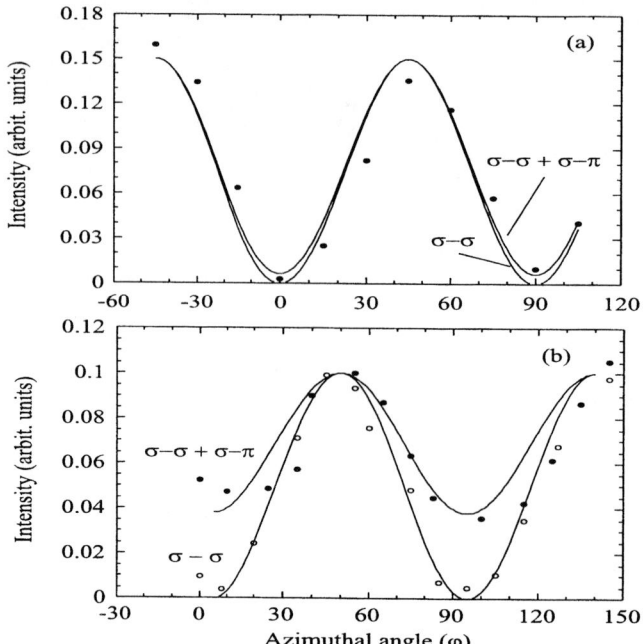

FIGURE 3. Intensities of the (002) and (006) forbidden reflection as a function of the azimuthal angle φ at E= 7.124 eV. The total intensity and the σ-σ channel are shown. Continuous line shows theoretical behaviour.

with an excited state **b**, is given by [9]

$$A \sim \frac{<a\mid \vec{e'}\cdot\vec{r}exp(-i\vec{k'}\cdot\vec{r})\mid b><b\mid \vec{e}\cdot\vec{r}exp(i\vec{k}\cdot\vec{r})\mid a>}{E_a - E_b + \hbar\omega - i\Gamma/2} \quad (1)$$

Expanding the exponential factors the previous expression is given by:

$$f(\vec{e},\vec{e'}) = Pf_0 + Q_d - iQ_{dq} + Q_q + \cdots \quad (2)$$

where, f_0 is the Thompson scattering, Q_d is a tensor of range 2, describing the dipolar transitions, Q_{dq} is the mix dipole-quadrupole term and Q_q is the pure quadrupolar term. Neglecting tensors of range higher than four, the structure factor for a crystal with N atoms per unit cell is:

$$F_{\mathbf{h}} = \sum_{n=1,N}[Pf_{0n} + Q_{dn} - iQ_{dqn} + Q_{qn}]D_n exp(i\vec{h}\cdot\vec{x_n}) \quad (3)$$

In our case the structure factor for the forbidden reflection is given by $F(002) = 4(f_{B1}+f_{B2}-f_{B3}-f_{B4})$ where B_i represent the octahedral iron ions within an octant of the crystallographic cell. A similar expression is obtained for the tetrahedral Fe ions. The site symmetry for the A atoms is T_d so both, the dipolar and quadrupolar anomalous scattering factors are isotropic and we can not expect any contribution to the structure factor arising from these two terms. Only the mixed dipolar-quadrupolar term is anysotropic and can give a possible contribution. For the B atoms, the site symmetry is trigonal ($\bar{3}m$). In this case, the dipolar scattering factor is anysotropic being the anysotropy axis in the direction of the threefold axis. Therefore the dipolar structure factor related to the B atoms is non-zero and the scattering intensity depends on the relative direction between the anysotropy axis and the polarization of the incident beam.

The dipolar contribution to the anomalous scattering tensor of each B atom is given, in the trigonal reference frame, by a diagonal tensor with two degenerate components (i.e. the two directions perpendicular to the threefold axis) and a third component in the direction of the trigonal axis. Upon transformation to the reference frame of the crystal ($\hat{x}$, $\hat{y}$, $\hat{z}$), the structure factor of these reflections is given by the following tensor:

$$F_{(004n+2)} = 16 \begin{pmatrix} 0 & f_b & 0 \\ f_b & 0 & 0 \\ 0 & 0 & 0 \end{pmatrix} \quad (4)$$

where $f_b = \frac{1}{3}(f_\| - f_\perp)$, being $f_\|$ and $f_\perp$ the anomalous scattering factors along and perpendicular to the trigonal axis, respectively, as reported by Dmitrienko [6]. The intensity of these ATS reflections is given by:

$$I_{\alpha\beta} = \left| \beta^+ F_{hkl} \alpha \right|^2 \quad (5)$$

where α and β are the unity vectors of the beam polarization for the incident and scattered light. In our experiments, the incident beam is σ polarized ($\vec{e}$ perpendicular to the difracction plane) and σ and $\sigma + \pi$ polarizations are measured for the scattered beam. The expressions for the intensity as a function of the azimuthal angle (φ) are the followings:

$$I^{\sigma\sigma'}_{004n+2} = |16 f_b|^2 \sin^2(2\varphi) \quad (6)$$

$$I^{\sigma\pi'}_{004n+2} = |16 f_b|^2 \sin^2(\theta) \cos^2(2\varphi) \quad (7)$$

$$I^{Tot}_{004n+2} = |16 f_b|^2 \left(\sin^2(\theta) + \cos^2(\theta) \sin^2(2\varphi) \right) \quad (8)$$

The experimental azimuthal dependence of the scattered intensity agrees nicely with the dipolar model, not only at the main resonance and at the extended part of the spectrum but also for the resonance at the prepeak. The difference between (002) and (006) intensities at the prepeak resonance can not be explained only by a dipolar transition and, as Eq. (1) shows, it also depends on the transfer moment ($\vec{k'} - \vec{k}$).

"Ab initio" multiple scattering calculations performed for a cluster around the tetrahedral and octahedral atoms have shown that the prepeak anomaly is due to a mixed dipole-quadrupole contribution from the A type atoms, being the contribution of B atoms zero. The main resonance and the extended part are due mainly to dipolar contribution of the B atoms. The comparison between experimental intensities and "ab initio" calculations for (002) and (006) reflections is shown in fig. 4.

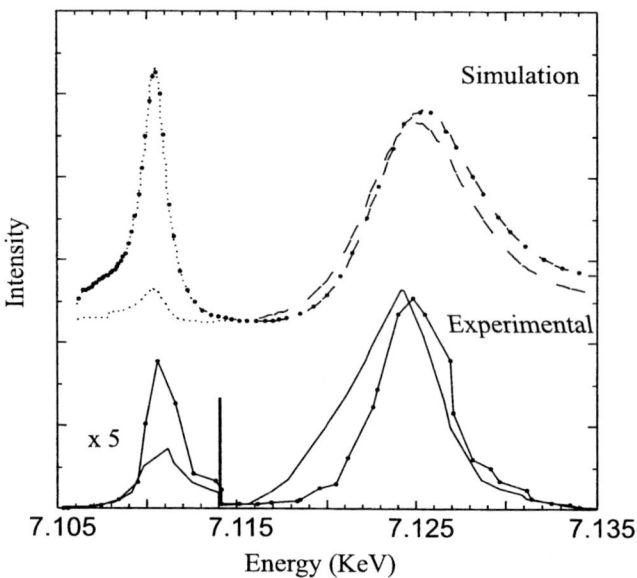

FIGURE 4. Comparison of the experimental energy dependence with the "ab initio" calculations of the scattered intensity for (002)(line) and (006) (dots)reflections. The main peak at 7124 eV is obtained from dipolar transitions at the octahedral iron atoms. The resonance at the prepeak energy, 7.11 KeV, is obtained when including dipolar-quadrupolar processes at the tetrahedral iron atoms.

In other words, the resonance at the absorption edge is due to the ordering of splitted B projected empty p orbitals. This splitting comes from the trigonal anysotropy of the B atoms. "Ab initio" multiple scattering calculation for the trigonal FeO_6 cluster gives a splitting of about 1 eV between the absorption coefficient for polarizations parallel and perpendicular to the threefold axis. The oscillations above the absorption edge have the same origin. It can be shown that this oscillatory signal comes from the addition of cosine terms whose frequencies are twice the difference between interatomic distances of different coordination shells.

$$I_{004n+2} \propto (f''_{0Fe})^2 \sum_i \Delta A_i^2 + (f''_{0Fe})^2 \sum_{i>j} 2\Delta A_i \Delta A_j \cos(2k(r_i - r_j) + (\delta_i - \delta_j)) \quad (9)$$

Summarizing, the observation of these forbidden reflections, its azimuthal (φ) and angular (θ) behaviour and the evolution with the energy of the incoming photons can be well explained by the anysotropy of the atomic anomalous scattering factor. The resonance at the prepeak is due to a mix dipolar-quadrupolar term of the tetrahedral A atoms, whereas the main resonance and the extended part are due to the dipolar contribution of the octahedral B atoms. It is important to underline that the anomalous scattering factor is the same for all the iron atoms, tetrahedral and octahedral. The only difference is the different orientation of their anysotropic axes in the unit cell.

Low temperature data ($T < T_v$)

The comparison of the energy dependence of the intensity of (002) and (006) forbidden reflections, above (room temperature) and below (30 K) the Verwey transition is shown in fig.5. No appreciable changes are observed, as it is also shown, in more detail, measuring the scattered intensity as a function of temperature and azimuthal angle [10] .

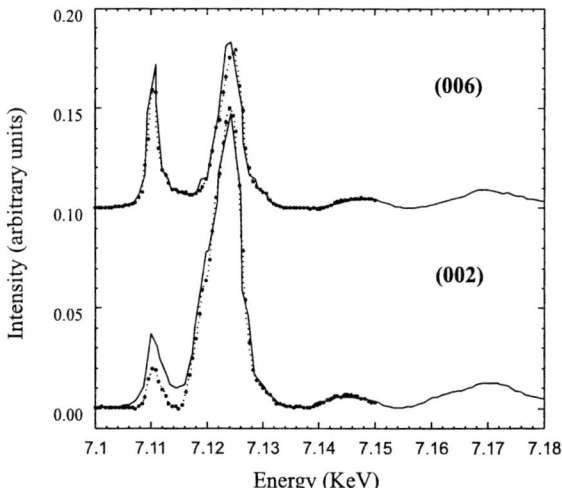

FIGURE 5. Intensity versus photon energy of (002) and (006) forbidden reflections with the incident light polarization vector parallel to the [110] crystallographic direction at room temperature (continuous line) and at 30 K (dotted line). Intensity is no selfabsorption corrected.

We have also performed experiments by cooling the sample with an external magnetic field applied. The comparison between the spectrum of the non-oriented sample (obtained cooling down a demagnetized sample without a magnetic field) and the partially oriented sample (cooling down under a field of 0.2 T) is shown in fig. 6 at temperatures below the Verwey transition.

The spectra are almost identical showing that the effect of crystal orientation does not have any influence on the scattered intensity. From the low temperature data

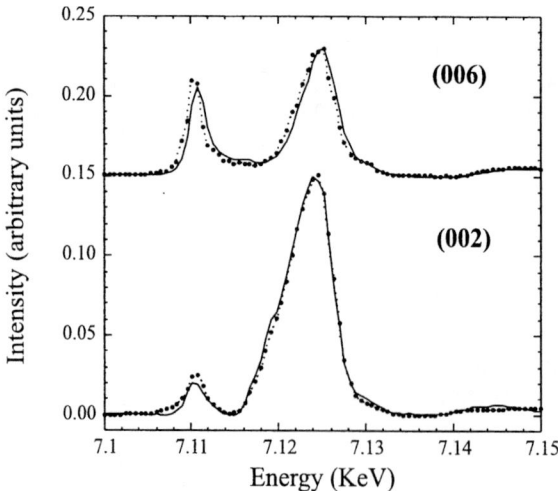

FIGURE 6. Comparison of the intensity of (002) and (006) forbidden reflections versus photon energy for the non-oriented (line) and partial oriented (dots) sample at 30 K. Intensity is no selfabsorption corrected.

we can conclude that the metal-insulator phase transition in Fe_3O_4 does not induce any change in the p and p-d orbital ordering.

DISCUSSION AND CONCLUSIONS

These experiments have a fundamental importance concerning the description of the electronic transport properties of magnetite. As it is well known, the absorption coefficient for compounds where atoms are present in different oxidation states, differs in their chemical shift, i.e. the absorption edge shifts at high energies with increasing the valence state. The Fe absorption edge in Fe^{3+} oxides is about 3-5 eV higher than in Fe^{2+} oxides. Therefore, the main difference between the anomalous scattering factors at the edge will arise from atoms with different valence states. In our case, the observed main resonance is due to the splitting of the absorption edge related to the charge anysotropy around the absorbing (scatterer) atom. The anysotropy splitting is about 1 eV that is quite lower than the 3-5 eV of the chemical shift. When charge ordering occurs, resonant scattering in forbidden reflections would be observed because of the appearance of a new periodicity, but the intensity of these reflections should not depend on the azimuthal angle.

The other point to consider is that the time involved in the virtual absorption process associated to the anomalous scattering atomic factor, is about 10^{-16} sec. So, the anomalous scattering diffraction probes the ordering on a very short time scale. The results show that only one kind of tetrahedral and octahedral iron ions exists on magnetite below and above the Verwey transition. The azimuthal behaviour of (002) and (006) reflections shows clearly that the appearance of these reflections is due to the local anisotropy of the tetrahedral atoms (mixed dipolar-cuadrupolar anisotropy) which is manifested in the prepeak resonance and of the octahedral atoms ($\bar{3}m$ local symmetry) which gives the main peak resonance, mainly of dipolar character. We will focus on the octahedral atoms, for which charge fluctuation and charge ordering mechanisms have been proposed. The mechanism for conduction in the high temperature phase is described in terms of charge fluctuation between octahedral iron ions. If the anomalous scattering factor depends on the valence state of the atom and the charge fluctuates in time scale lower than 10^{-16} sec, the incident photon should see a random distribution of different scattering factors, due to Fe^{3+} and Fe^{2+} atoms and no coherence for diffraction would be fulfilled. We have also considered the case, where the difference in the atomic scattering factor is only due to the chemical shift and the anisotropy is the same for Fe^{3+} and Fe^{2+}. The shape of the main resonance as a function of the energy would be double peaked in this case instead of the single peak observed. Therefore, we can conclude that the charge does not fluctuate in the high temperature phase of magnetite and the conduction should be explained in terms of band structure. In the insulating phase, at temperatures below the Verwey transition, the charge ordering would give an isotropic contribution at the absorption edge resonance. Since the spectra do not change above and below the phase transition and between non-oriented and partial oriented sample and they show the same azimuthal behaviour at room and at low temperatures, we can conclude that no charge ordering occurs below the phase transition. A detailed discussion of these points are given in refs. 8 and 10. Summarizing, resonant scattering of (002) and (006) reflections demonstrates the absence of either temporal or spatial charge localization in magnetite.

REFERENCES

1. Tsuda N., Nasu K., Yanase A., and Siratori K., *Electronic conduction in oxides*, Springer, Heidelberg, 1991.
2. Robin M.B., and Day P., *Adv.Inorg. Radiochem.* **10**, 247 (1967).
3. Murakami, Y. et al, *Phys. Rev. Lett.* **80**, 1932 (1998).
4. Kanamori, J. and Kotani, eds, *Core level spectroscopy in Condensed Systems*, Springer, Heidelberg, 1988.
5. Coppens Ph., *Synchrotron Radiation Crystallography*, Academic Press, San Diego, 1992.
6. Dmitrienko, V. E., *Acta Cryst.* **A39**, 29 (1983). *Acta Cryst.* **A40**, 89 (1984). Templenton, D.H. and Templenton L.K., *Acta Cryst.* **A41**, 133 (1985). *Acta Cryst.* **A42**, 478 (1986).
7. Verwey E. J. W., *Nature (London)* **144**, 327 (1939).
8. Garcia, J. et al., *Phys. Rev. Lett.*, submmited.
9. Templenton D. H., and Templenton L. K., *Phys. Rev. B* **49**, 14850 (1994).
10. Garcia, J. et al., to be published.

Orbital ordering and resonant diffraction in manganites

Yves Joly[1], Maurizio Benfatto[2], and Calogero R. Natoli[1,2]

[1]*Laboratoire de Cristallographie, CNRS, associé à l'Université Joseph Fourier, B.P. 166, F-38042 Grenoble Cedex 9*
[2]*Laboratori Nazionali di Frascati, INFN, Casella Postale 13, I-00044 Frascati*

Abstract. The experimental evidence for orbital ordering (OO) in $LaMnO_3$ is critically reexamined in the light of realistic calculations based on finite difference method (FDM) for the resonant x-ray scattered intensity at a forbidden reflection. After some introduction on the way of calculation, it is shown that the main contribution to the scattered intensity is not due to orbital ordering but to the Jahn-Teller distortion. It is concluded that the dipole resonant x-ray diffraction at the Mn K-edge is not a suitable technique for the *direct* observation of the orbital ordering in the considered compounds. A short glance on the Fe_3O_4 compound is also given demonstrating a mixed dipolar-quadrupolar contribution in the anomalous signal.

INTRODUCTION

Transition metal oxides offer a wide varieties of electric and magnetic properties. The interplay between the charge, spin and orbital degrees of freedom is the key of the understanding of the origin of magnetism and the nature of the electronic states [1,2]. These traditional topics have recently known a resurgence of interest because of the discovery of a new phenomenon of fundamental and technological importance which was referred to as colossal magnetoresistance in perovskite-type manganites.

In the same time, the second and third generation synchrotron radiation sources allows experiments which were impossible in the past. In particular, while the charge and spin ordering could be investigated by the traditional neutron and electron diffraction techniques, the *direct* experimental observation of the orbital ordering, predicted on the basis of theoretical considerations, has until very recently proved quite difficult. In the last few months however experimental works, using the resonant x-ray scattering technique at transition metal K-edge around forbidden reflections have reported observation of scattered intensity which has been presented as providing *direct* experimental evidence of the theoretically predicted orbital ordering in these materials. In the crystals under investigation [3,4], respectively in $La_{0.5}Sr_{1.5}MnO_4$ and $LaMnO_3$ perovskite(-type) compounds, the signal was associated with a flipping of the polarization plane of the incident radiation. An even more recent experimental observation of the orbital ordering was reported by Paolasini *et al* [5] concerning V_2O_3.

Such new experimental developments need new theoretical approaches to interpret quantitatively the experiments and to get a safe basis for the understanding of the observed signals. A new scheme of calculation, avoiding classical approximations on the shape of the potential, and already used in X-ray absorption near edge structure (XANES) is extended to anomalous diffraction. This way of calculation based on the finite difference method (FDM) is briefly presented in the next section. Applied to the $LaMnO_3$ perovskite compound, it will be shown that the observed signal is due to the Jahn-Teller (JT) distortion and not to the orbital ordering. In the last section, to continue on the fact that care must be taken in the interpretation of anomalous signal, a complementary glance is given on the Fe_3O_4 magnetite.

CALCULATION OF RESONANT DIFFRACTION

Generalities

A number of spectroscopies are related to the transition process of a core electron to some unoccupied level. Due to the transition selection rules, these spectroscopies authorizes selective analysis of the valence state around a chemical selected specie. The success of XANES in all classes of material demonstrates the powerness of this way of investigation.

The mechanism wereby x-rays are resonantly scattered is well known. The incoming photon is virtually absorbed to promote a core electron to an empty intermediate excited state, which subsequently decays to the same core hole emitting a second photon with the same energy but eventually with another polarization than the incoming one. Because the diffraction process is coherent, one adds to the chemical and moment selection processes, the site selection process. The total atomic factor is the sum of the scattering amplitudes coming from all the resonant atoms of the unit cell multiplied by a site dependent phase factor. For a forbidden reflection, the atomic scattering amplitudes just depend on the atomic number and for specific beams, the phase factor makes that the contributions coming from the different atoms subtract each other and the total scattering amplitude is zero. Nevertheless at a threshold, the resonant process depend on the transition probability for the photoelectron to transite to the empty states. These ones depend strongly to *any* anisotropy of the environment (be it of structural origin or due to the anisotropic charge distribution in the ground state of the system) it gives rise to a tensor component in the atomic scattering factor (ASF) with sharp photon energy dependence in the anomalous dispersion region. This anisotropy of the x-ray susceptibility tensor leads to a number of well known phenomena, like pleochrism, existence of " structurally forbidden " reflections [6,7], birifringence, orientational dependence of X-ray Absorption Near Edge Structure (XANES) and reflectivity, etc..

One can understand that a confident calculation must include a precise evaluation of these intermediary states that the photoelectron comes to sound. This evaluation must take into account all the region around the absorber which can be responsible of the anisotropy. For this purpose and because we want to check in this study the effect of the orbital ordering on the signal we have to use a way of calculation which is sensitive to it. The finite difference method is among other, see for instance the work of Huhne , Ebert and coworkers [26,27], a good way to solve the Schrödinger equation in a free shape potential. Such approach is necessary when one wants to check the influence of the orbital ordering. Indeed, the associated potential cannot be described in the spherically averaged potential used in most multiple scattering calculation. That method and its inclusion in anomalous diffraction calculation is presented in the following.

The finite difference method

The method has been known for a long time [12] but it is only recently that the first FDM band structure calculation was reported [13]. The technique was also extended to Low Energy Electron [14] and Positron Diffraction [15]. Sensitivity of low energy particles on electronic parameters was illustrated in Ref. [16]. The general way to apply the FDM to XANES was published very recently [18]. The first study on anomalous scattering was reported some months later [19].

In x-ray absorption spectroscopies the signal is calculated from the construction of matrix transition $< \psi|o|g >$ connecting the core state g to final states ψ, through a transition operator o which will be shown further on. The FDM is used to calculate these final states ψ, solving in direct space the Schrödinger equation.

The calculation is performed on a sufficiently large cluster around the central atom. Following Dill and Dehmer [17], the whole space is divided in three regions: i) an outer sphere surrounding the cluster of interest, ii) an atomic region composed by very little spheres (up to 0.65 Å of radius) around the atomic cores and not the usual large "muffin-tin" spheres, and iii) the inter-atomic region where an FDM formulation of the Schrödinger equation is performed. In the outer sphere region, the potential is assumed constant and a complete set of solutions is given by an expansion of the final states in spherical harmonics with $L = (\ell, m)$:

$$\psi^{L,E}(\mathbf{r}) = \sqrt{\frac{\kappa}{\pi}} \left(j_\ell(\kappa r) Y_L(\hat{\mathbf{r}}) + \sum_{L'} \tau_{L'}^{L,E} h_{\ell'}^+(\kappa r) Y_{L'}(\hat{\mathbf{r}}) \right),$$

where j_l and h_l^+ are the radial Bessel and Hankel outgoing functions. $\kappa = \sqrt{E}$ is the electronic wave vector and E is the kinetic energy of the photoelectron in the outer sphere. The quantities $\tau_{L'}^{L,E}$ are the unknown cluster scattering amplitudes to be determined in the way specified below. The factor $\sqrt{\frac{\kappa}{\pi}}$ assure normalization of the scattering wave function to one state per Rydberg.

In the atomic core region, since the potential is spherically symmetric to a very good approximation, one can expend the solution as :

$$\psi^{L,E}(\mathbf{r}) = \sum_{L'} A_{L'}^{L,E} R_{\ell'}^{E}(r) Y_{L'}(\hat{\mathbf{r}}),$$

where the functions $R_{\ell'}^{E}(r)$ are solutions of the radial Schrödinger equation inside the atoms, with the vector $\mathbf{r}$ being measured from the atomic centers.

In the interatomic region, the unknowns are the values of the wave function on each grid point i: $\psi_i^{L,E} = \psi^{L,E}(\mathbf{r}_i)$. Here, the Laplacian is obtained by approximating the wave function around the point i by a polynomial of order four. In this way, in an orthogonal frame the Schrödinger equation becomes:

$$[6 + h^2(V_i - E)]\psi_i^{L,E} - \frac{16}{15}\sum_{j}^{\text{first}} \psi_j^{L,E} + \frac{1}{15}\sum_{j}^{\text{second}} \psi_j^{L,E} = 0,$$

expressed in Rydberg and atomic units. h is the interpoint distance; the smaller it is, the more accurate is the computation. $V_i = V(\mathbf{r}_i)$ is the potential on the node point i. The summations are respectively over the six first and the six second neighbor points in the three orthogonal directions.

The amplitudes of the spherical harmonics within the atomic core region and the outer sphere are evaluated using the continuity of the wave function and of its derivative between the different regions. Finally a large system of linear equations is obtained connecting the values of the wave function on all the grid points.

This formulation does not require any approximation on the shape of the potential. This latter, moreover, to a very good approximation can be assumed to be a functional of the total electronic density relaxed around the core-hole. The total electronic density is described by the superposition of the charge densities of the atoms whether neutral or ionized according to the system. In particular bond directed valence orbitals are used whenever appropriate what is the case in the present study. In $LaMnO_3$, where the effect on the signal of the orbital ordering must be checked, the electron densities is evaluated in that way, populating a specific d orbital in a specific direction. Then the coulomb potential is calculated solving Poisson equation and the exchange-correlation potential is calculated following Hedin-Lundqvist. Therefore the shape of a spectrum is determined not only by the position of the atoms around the photo absorber but also, although to a lesser extent, by the electronic density of the system.

First, to prove the capacity of the method to give precise results, one presents as a test case a comparative calculation of the XANES Cu K-edge, in a little cluster having the central copper plus its 12 surrounding first neighbors. The comparison is done between our approach and the classical multiple scattering frame. In order to have the best comparison, exactly the same potential is used,

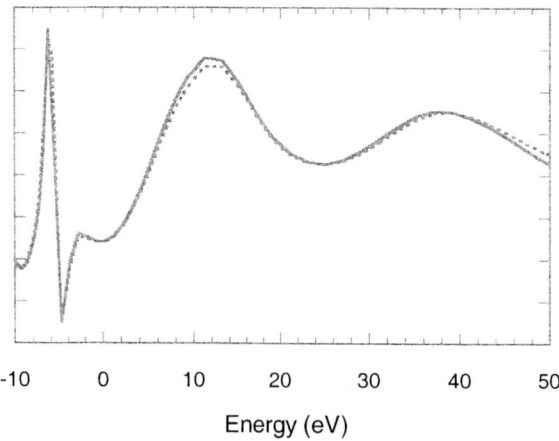

FIGURE 1. Comparison between the FDM (full line) and MST (dotted line) calculations of the XANES at the copper K-edge in a 13 copper atoms cluster. The occupied states are not eliminated

a muffin-tin one. Anyway, for metallic copper, which is a dense metal, this is a very good approximation. In the MST calculation the multiple scattering matrix is inverted and not expanded in path. The number of spherical harmonics is governed by the classical law giving the maximum momentum number : $\ell_{max} = \kappa R_{mt}$, R_{mt}, being the muffin-tin radius. As can be checked in Fig. (1), the agreement is quite perfect with an interpoint distance of $h = 0.25 \text{Å}$. If one wants to make calculation at higher energy a denser grid of point is necessary. In real XANES study the approach already proved to be efficient to analyze charge exchange and core-hole screening effect in rutile TiO_2 [18].

Inclusion in resonant diffraction

Having calculated the intermediary states $\psi^{L,E}$, each matrix transition can be easily evaluated. Then, the scattering amplitude resulting from the resonant absorption-emission of a photon by an atom is given by :

$$f = (\hbar\omega)^2 \sum_{L,E} \frac{<g|o_s^*|\psi^{L,E}><\psi^{L,E}|o_i|g>}{\hbar\omega - (E_\psi - E_g) + i\frac{\Gamma}{2}} \quad (1)$$

where E_g and E_ψ are the initial and intermediary state energies. In a one body approach, E_ψ is linearly related to the photoelectron kinetic energy E by : $E_\psi = (E + I_g)$, where I_g is the ionization energy of the g level, for instance the $1s$ one, up to the Fermi level E_f. $\hbar\omega$ is the photon energy. o_s and o_i are the transition operators, the indexes s and i standing for the outcoming and incoming photons. At second order they are given by :

$$o_{s,i} = \epsilon_{s,i}.\mathbf{r}(1 + \frac{i}{2}\mathbf{k}_{s,i}.\mathbf{r}) \qquad (2)$$

where ϵ is the polarization, and $\mathbf{k}$ the photon wave vector. To get the total amplitude scattered by one unit mesh, the summation over the atoms a, with their phase factor $e^{i\phi_a}$ must be included. The summation over the final states is splitted between states having the same energy. In dipolar approximation and using the tensor approach, one gets :

$$f = \sum_a e^{i\phi_a} \sum_{\alpha\beta} \epsilon_\alpha^i \epsilon_\beta^s f_{\alpha\beta}^a(\omega)$$

with for each atom and omitting the index a :

$$f_{\alpha\beta}(\omega) = (\hbar\omega)^2 \int_{E_f}^\infty dE \sum_L \frac{<g|r_\alpha|\psi^{L,E}><\psi^{L,E}|r_\beta|g>}{\hbar\omega - I_g + E_g - E + i\frac{\Gamma(E)}{2}}$$

where we recall that $\psi^{L,E}$ is the continuum scattering solution of the Schrödinger equation for the cluster of atoms at energy E above the Fermi level in response to an exciting wave $J_L(\vec{r})$ calculated using the real part of the Hedin-Lundqvist self-energy. $\frac{\Gamma(E)}{2}$ is then the sum of the imaginary part of this latter plus the half-width at half-maximum of the core hole state.

APPLICATION TO MANGANITE

Structure

The $LaMnO_3$ space group is Pnma. The unit cell (Fig. (2)) contains four manganese surrounded by distorted oxygen octahedron. Following the crystal structure established by Norby et al. [21], the rhombohedral unit mesh parameters are $5.5392 Å$, $5.6991 Å$ and $7.7175 Å$. The MnO distances are $1.9169 Å$, $1.9647 Å$ for the shortest ones, and $2.1446 Å$ for the longest one. Indeed, the Mn^{3+} in $LaMnO_3$ is a typical Jahn-Teller ion with the electron configuration in octahedral coordination of $t_{2g}^3 e_g^1$. The specific orbital ordering with an oriented occupied d orbital is given in the same figure; the conjugated ferroelectric spin order in basal plane due to the superexchange is also shown. On the contrary the exchange interaction along the c direction is antiferromagnetic. This system is rather close to a cubic one shown in thick line. The cube, as in the Murakami et al paper [3], is often used as reference. The remaining lanthanum atoms resides at the center of these cubes.

Keeping the rhombohedral unit cell the four Mn atoms are crystallographically equivalent after π rotations around the a, b and c directions. Taking the atom, say Mn_1, at the corner as reference, the other basal plane atom (Mn_2), $(\frac{1}{2}, \frac{1}{2}, 0)$ is equivalent after the rotation around b, for the $(0, \frac{1}{2}, \frac{1}{2})$ and $(\frac{1}{2}, \frac{1}{2}, \frac{1}{2})$ Mn atoms, the rotation are respectively around the c and a directions. If the system is approximated in the cubic cell (not done in the complete calculations) the two basal

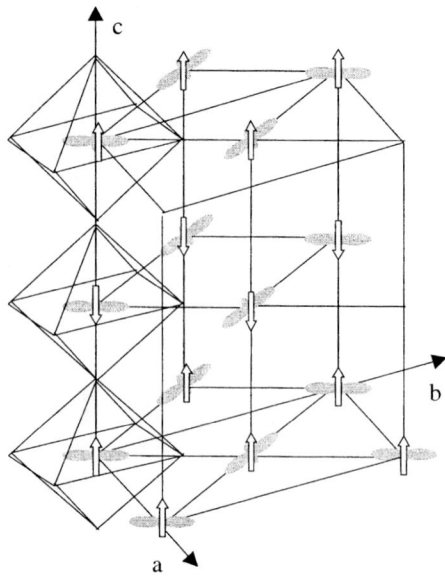

FIGURE 2. The LaMnO$_3$ structure. The rhombohedral unit mesh is shown together with the simplified cubic lattice (thick line). The orbital ordering shown on the figure is responsible of the spin ordering represented by the arrows.

plane atoms are equivalent after a $\frac{\pi}{4}$ rotation along the c axes. The two other atoms becoming simply equivalent without rotation to the basal plane ones.

Previous results

So, in the particular case under consideration [4], Murakami *et al* have reported the observation of scattered intensity, at the forbidden (3,0,0) reflection, with flipping of the polarization plane of the incident radiation. The explanation is given in the cubic simplified mesh : the excitation of the Mn 1s electron to empty 4p states (splitted in their y and x, z components by an energy Δ due to some interaction with the surrounding, to be specified in the following) gives rise to non zero resonant scattered intensity proportional to Δ^2 at the forbidden reflections (3,0,0), which is sensitive to the difference between the ASF of the two "orbitally ordered" sublattices. The origin of the splitting Δ is not specified in the model proposed in ref. [4] but the authors clearly state that one possible source is the Coulomb interaction between the 4p conduction band states and the ordered 3d orbitals. An alternative mechanism, they suggest, comes from the coherent Jahn-Teller distorsion (JTD) of the oxygen octahedra surrounding the Mn atoms that accompanies the orbital ordering, with the long axis always along the occupied $3de_g$ orbital.

Now this same model (which we believe substantially correct), leads to the conclusion that the resonant diffracted intensity due to the two effects is at least two orders of magnitude greater in the case of the JTD. In fact a simple XANES calculation of the x, y and z absorption components of the $1s$ to $4p$ dipole transition of a Mn atom surrounded by six oxygens at their crystallographic positions shows a splitting of the x and y components of 2.0 eV, whereas the Coulomb splitting of the $4p_{x(z)}$ and $4p_y$ Mn orbitals turns out to be $\frac{6}{35}F^2 = 0.4 eV$, taking for F^2, the Coulomb Slater integral for the $3d$ and $4p$ atomic Mn orbitals, the value of 2.5 eV as calculated in the atomic approximation using Cowan program [8]. This already implies a factor of 25 in favor of the JTD mechanism and certainly represents a lower bound, since screening and band effects in the solid tend to drastically lower the value of F^2. A not unreasonable further reduction of this latter by a factor of two provides the two orders of magnitude difference in the diffracted intensities. Notice, by the way, that Ishihara et al [9] in their model calculations on the anomalous x-ray scattering in manganites had to assume for F^2 a value of 13 eV to get a Coulomb splitting of 1.8 eV between the $4p_z$ and $4p_{x(y)}$ orbitals (they use a rotated frame), leading to a 1.3 eV splitting of the corresponding peaks of the imaginary part of the scattering factor. However such an high value for F^2 is clearly unreasonable.

Then, one has to substantiate these considerations. Since the point group symmetry at the Mn site is the same for JTD and orbital order, "theoretical" experiments are needed to disentangle the different contributions and assess the relative intensity originating from the two competing mechanisms. We performed realistic model calculations for the ASF using both FDM and multiple scattering approaches.

Symmetry of the experiment

Already keeping the cubic simplified scheme, the symmetry operation explained above give that $f_{xx}^{Mn_1} = f_{yy}^{Mn_2}$ and $f_{zz}^{Mn_1} = f_{zz}^{Mn_2}$. Consequently the anomalous scattering amplitude at the (300) "forbidden" reflection, can be written as

$$f = 2(\epsilon_x^i \epsilon_x^s - \epsilon_y^i \epsilon_y^s)(f_{xx}^{Mn_1} - f_{yy}^{Mn_1}) \qquad (3)$$

In the conditions of the experiment of ref. [4] and indicating by ψ the azimuthal angle around the scattering vector, we have for the $\sigma \to \pi$ channel, $\vec{\epsilon}^i = (\sin\psi/\sqrt{2}, -\sin\psi/\sqrt{2}, \cos\psi)$ and $\vec{\epsilon}^s = (\sin\theta - \cos\theta\cos\psi)/\sqrt{2}, (\sin\theta + \cos\theta\cos\psi)/\sqrt{2}, \cos\theta\sin\psi)$, where $\theta = 30°$ is the Bragg's angle, it is immediately seen from Eq. (3) that the scattering amplitude is proportional to $\sin\theta\sin\psi$, providing therefore the observed intensity dependence on the azimuthal angle ψ. Clearly the intensity for the $\sigma \to \sigma$ channel is zero since then $\vec{\epsilon}^i \equiv \vec{\epsilon}^s$.

Multiple scattering calculations

To assess the effect of the *sole* JTD we first perform calculation without any orbital ordering. In that case the muffin-tin approximation must be sufficient. Even

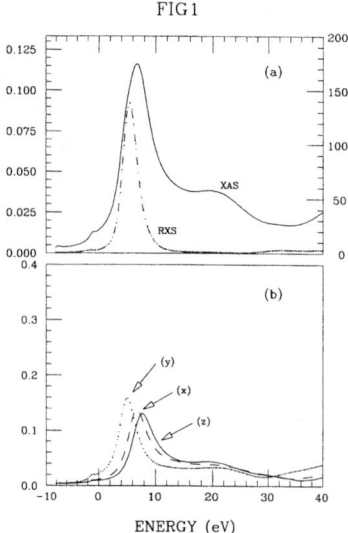

FIGURE 3. Panel (a): Mn K-edge unpolarized absorption cross section in LaMnO$_3$ and anomalous resonant diffraction intensity in the muffin-tin approximation; Panel (b): polarized cross sections along the principal axis.

if the paper is principally devoted to the FDM formalism, at least for comparison, we performed also calculations in the frame of multiple scattering theory (MST) [11]. As expected, and as it was already checked in XANES when the potential is supposed to be close to a muffin-tin one, both approaches give qualitatively the same results.

In the MST context, following Vedrinskii *et al* [20] one can write the adimensional anomalous part of the ASF anisotropic tensor at the photon energy $\hbar\omega$, using atomic units, as

$$f_{\alpha\beta}(\omega) = (\hbar\omega)^2 < 1s|\vec{r}_\alpha G(\vec{r},\vec{r}';E_c)\vec{r}_\beta{'}|1s> \qquad (4)$$

Here $E_c = \hbar\omega + E_{1s} + i\frac{\Gamma}{2}$ is the complex excitation energy and $G(\vec{r},\vec{r}';E_c)$ is the Green function of the system. $\Gamma/2$ is defined in the same way than in the FDM formulation seen above. For more details we refer to ref. [11,22]. In Eq. (4) we have neglected the smooth background contribution coming from the integration over the occupied states [20]. That effect is responsible of the residual differences between the MST and the FDM approaches.

The calculation presented in Fig. (3) is performed on a 21 atom cluster around a central Mn with a radius of 5 Å. It contains 7 Mn atoms with a distorted oxygen octahedron (6 O) and 8 La atoms, enough to reach cluster size convergence due to the finite mean free path of the excited electron.

The upper panel (a) in Fig. (3) shows the first 40 eV of the unpolarized absorption cross section in Mbarn (0.11 at peak height) together with the resonant diffraction signal (the square of Eq. 3 with $\sin\theta = 1/2$ and $\psi = \pi/2$, 250 at the maximum, corresponding to a number of effective scattering electrons of the order of five). The lower panel (b) displays instead the anisotropy of the polarized absorption along the principal axis which is responsible for the effect. The main experimental peak is well reproduced. Clearly the intensity of the resonant signal is related to the linear dichroism of the crystal along the x and y directions. Notice that no charge order is present in the model, since the charge is spherically symmetric around each atom. Substantially the same result is obtained from a 51 atom cluster calculation containing 7 Mn atoms with their distorted oxygen octahedra (36 O) and 8 La atom, with a radius of 6 Å, thus confirming cluster size convergence.

Finite difference method calculations

We then performed non-muffin-tin calculations using the FDM approach. In the absence of a self-consistent charge density with orbital order we used an input charge density obtained from the superposition of neutral atomic charge densities with a Mn configuration given by $t_{2g}^3 e_g^1 4s^2$ with the last e_g electron oriented in the right way. From this charge density we constructed the Coulomb potential of the cluster and the Hedin-Lundqvist self-energy.

In order to assess directly the relative weight of the two competing mechanisms we preliminarily performed two seven atom calculations (a central Mn and six O, not shown), first in a distorted octahedral coordination without orbital ordering and then with a fictitious orbital ordering on the central Mn (no core hole relaxation) and no distortion. The calculated scattered intensities are in the ratio one to one hundred in favor of the JTD, in keeping with the rule of thumb estimate given above. A further calculation in which both distorsion and orbital ordering are present shows that there is a distructive interference between the two mechanisms, leading to a decrease of scattered intensity, compared to the case of pure distorsion, by roughly 20%, therefore confirming the intensity ratio between the calculation with only JTD and that with only OO. This is understakable since the elongation of a bond (eg in the y direction) tends to diminish the energy of the rising edge in the same polarization, whereas the orbital occupancy along the same direction tends to increase it, due to the repulsive Coulomb interaction.

We then performed three calculations in the 21 atom cluster mentioned above, assumed in their cristallographic positions, one with no OO (to serve as a reference), one with OO on all Mn atoms and the last one with OO only on neighboring Mn atoms. In this latter case for the photoabsorber Mn atom we adopted the "well screened" configuration $t_{2g}^3 e_g^2$ which, being of cubic symmetry, does not contribute to the resonant signal. We believe, at variance with the authors of ref. [9], that this configuration is predominant in the final state, since the absorption spectrum

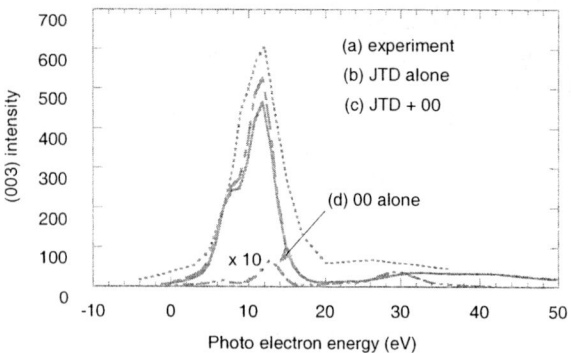

FIGURE 4. Anomalous resonant diffraction intensity at the (3,0,0) reflection by full potential calculation (FDM). a) dotted line: experiment, b) long-dashed line: pure JTD; c) full line: JTD plus OO on all Mn atoms; d) short dashed line: OO alone multiplied by 10

does not show sign of the presence of two distinct configurations in the final state, as in other systems [10]. A fourth calculation was performed keeping the OO but using non distorted octahedra to check the the effect of the pure orbital ordering. The main feature in the absorption coefficient in the JTD calculation is similar to the one already calculated in the muffin-tin case. Fig. (4) shows the result. The full line shows the effect of the *sole* JTD. The long dashed line shows the result of the full interference between JTD and OO on all Mn atoms. The short dashed line show the effect of the *sole* OO. This curve must is multiplied by 10 in the figure. It is clear from these calculations that in the absence of an absolute calibration of the intensity due to the pure JTD it is impossible to assess the effect due to OO. Therefore the claim made in ref. [4] of a *direct* observation of OO in $LaMnO_3$ is not substantiated by our findings, in contrast with the conclusions of ref. [9]. On the other hand the temperature dependence of the observed intensity at the forbidden (3,0,0) reflection, which disappears at $T_O = 780$ K in concomitance to a structural phase transition, is well in keeping with our analysis. X-ray resonant scattering for transitions to empty $3d$ states would be more suitable for such a *direct* observation, as pointed out in ref. [23], since then the "well screened" configuration $3d^{n+1}\underline{L}$ would be much less effective due to the self-screening effect of the excited photoelectron.

Yet another mechanism concurs to depress the resonant signal due to orbital ordering. This effect originates from the electronic relaxation of the photoabsorbing Mn atom following the creation of the $1s$ core hole. It is known in fact that, except in particular cases [10] (fluctuating valence, etc..) which is not the present one however, the final state "well screened" configuration $t_{2g}^3 e_g^2 \underline{L}$, where $\underline{L}$ indicates a hole in the oxygen $2p$ band, carries almost 90 % of the weight and therefore, being of cubic symmetry, does not contribute to the signal. Only the far away Mn atoms can, but their signal is depressed by the limited mean free path of the photoelectron in the final state. Calculations with orbital ordering on the neighboring Mn but not on the central one confirms that the signal becomes completely negligible.

Comparison with other studies

Two other recent papers confirm our findings. The first one from Elfimov *et al* [30] uses a LSDA + U band structure calculation. The authors do not go up to the evaluation of the anomalous signal but they clearly show that the effect of the orbital ordering is negligible in comparison with the Jahn-Teller distortion. They also suggest that at lower energy a projection of the d unoccupied orbitals of the neighboring Mn gives a little contribution in the p central Mn band. A structure at this energy would be a signature of the *sole* OO. We agree in principle, the difficulty remaining in the fact that such a signal must be at least hundred time smaller than the one devoted to JTD. Is it possible to detect such a low signal in the tail of the main Jahn-Teller feature ?

The second paper from Takahashi *et al* [31] uses the same formalism but, the authors provide in addition to the band structure results, the forbidden diffracted intensities resulting from the different crystallographic models. They also find that the signal is mainly due to the JTD and roughly proportional to the square of the magnitude of the local distortion. The authors point another important fact, that is that the signal is insensitive to the magnetic order.

A GLANCE ON FE_3O_4

In a recent experiment on Fe_3O_4, J. Garcia and coworkers [28] have shown a remarkable oscillations of the anomalous signal of two forbidden reflections. At the Fe ionisation energy, they extend to a very wide energy range. Nevertheless we focussing on the first pikes at the threshold, a first structure, corresponding to the tetehedral iron pre-edge fluorecence spectra is present. This peak depends drastically on the incidence angle as can be checked comparing the (002) and (006) reflections (see Fig. (5)). Starting from the general equation giving the atomic scattering (Eq. (1)), but extending the calculation to the quadrupole (Eq. (2)), and taking θ as the incidence angle, it can be checked that the scattering amplitudes are respectively for the octahedral and tetragonal sites :

$$f_o^{\sigma\sigma} = 2\sin 2\phi \left(f_{d_x d_x} - f_{d_y d_y} + \sin^2\theta \left(f_{q_{xy}q_{xy}} - f_{q_{yz}q_{yz}} \right) \right) \quad (5)$$

and

$$f_t^{\sigma\sigma} = 4\sin 2\phi \sin\theta f_{d_x q_{xy}} \quad (6)$$

where the atomic scattering amplitudes $f_{d_x d_x}$ and $f_{d_y d_y}$ refer to the dipolar contributions with the incoming and outcoming photon polarizations along x in the

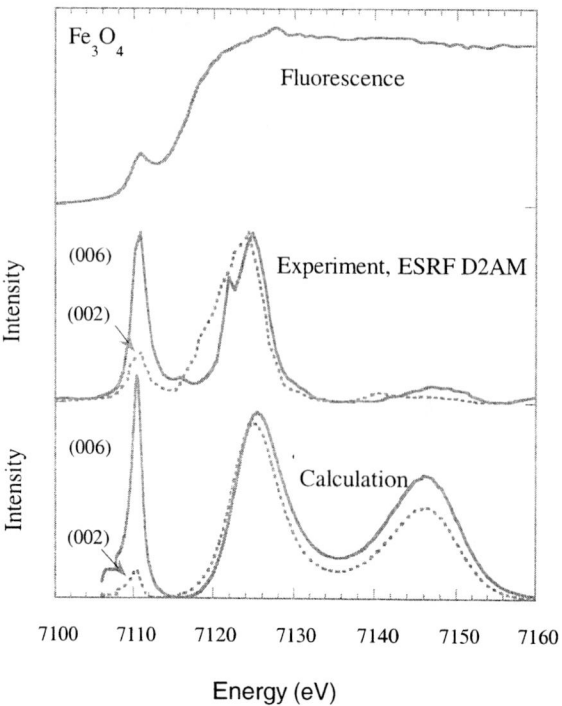

FIGURE 5. Anomalous resonant diffraction intensity at the (0,0,2) (dotted line) and (0,0,6) (full line). Top, the fluorescence spectrum with the pre-edge structure attributed to the tetragonal iron; middle, the experiment performed by J. Garcia and coworkers at the ESRF; bottom, first calculations

first case and y in the second; $f_{q_{xy}q_{xy}}$ and $f_{q_{yz}q_{yz}}$ are the quadrupolar contributions with polarization along y and the wave vector along z or x; $f_{d_x q_{xy}}$ is a hybridized dipolar-quadrupolar contribution where the incoming photon is absorbed through the dipolar channel, then the decay is through the quadrupolar channel with the polarization along x, emitting a new photon with the polarization along x and the wave vector along y. Such a phenomenon can occur only when the quadrupolar and dipolar transition matrix are non zero simultaneously, that is for non centrosymmetric site what is the case in the tetragonal iron site but not in the octahedral one. The scattered amplitudes coming from the two types of atoms interfere giving at the end the observed signal. The sinus dependence (to the square for the intensity) is roughly conform to the experiment. Some previous quantitative calculations confirm these results as can be checked in Fig.(5).

With that glance on Fe_3O_4, we just wanted to remind that this effect is possible in anomalous diffraction in rectilinear polarization, contrary to XANES where the same hybridized phenomenon can occur only with circular polarization (what is called natural dichroism). That fact was already revealed by Templeton and Templeton [29].

CONCLUSION

The main conclusion is that a lot of care must be taken in the interpretation of anomalous spectroscopy. The technique is at the same time very precise and very sensitive to a lot of parameters. In some cases it can become an excellent tool to analyze fundamental phenomena unreachable by other techniques. All the developments presented in that paper must be continued, in particular with the introduction of the magnetism with the spin-orbit coupling and to finish with the inclusion of correlated phenomena. With such tolls we hope to reach the level of the actual experimental data.

REFERENCES

1. Goodenough J.B., *Magnetism and Chemical Bond*, Interscience, New York, (1963)
2. Kugel K.I. and Khomskii D.I. Usp. Fiz. Nauk. **136**, 621 (1982) (Sov.Phys. Uspekhi **25**, 231 (1982))
3. Y. Murakami *et al.*, Phys. Rev. Lett. **80**, 1932 (1998).
4. Y. Murakami *et al.*, Phys. Rev. Lett. **81**, 582 (1998).
5. L. Paolasini *et al.*, Phys. Rev. Lett. **82**, 4719 (1999).
6. V.A. Belyakov, Usp. Fiz. Nauk. **115**, 533 (1975).
7. V.E. Dmitrienko, Acta Crystallogr. Sect. A **39**, 29 (1983); *ibidem* **40**, 89 (1984).
8. R.D. Cowan, *The Theory of Atomic Structure and Spectra*, University of California Press, Berkeley, Los Angeles, London (1981)
9. S. Ishihara and S. Maekawa, Phys. Rev. Lett. **80**, 3799 (1998).
10. Z.Y. Wu *et al.*, Phys. Rev. B **54**, 13409 (1996)
11. T.A. Tyson *et al.*, Phys. Rev. B **46**, 5997 (1992).
12. G.E. Kimball and G.H. Shortley, Phys. Rev. **45**, 815 (1934).
13. J.M. Thijssen and J.E. Inglesfield, Europhys. Lett. **27**, 65 (1994).
14. Y. Joly, Phys. Rev. Lett. **68**, 950 (1992).
15. Y. Joly, Phys. Rev. Lett. **72**, 392 (1994).
16. Y. Joly, Phys. Rev. B **53**, 13029 (1996).
17. D. Dill and J.L. Dehmer, J. Chem. Phys. **61**, 692 (1974).
18. Y. Joly *et al.*, Phys. Rev. Lett. **82**, 2398 (1999).
19. M. Benfatto *et al.*, Phys. Rev. Lett. **83**, 636 (1999).
20. R.V. Vedrinskii *et al.*, J. Phys.: Condens. Matter **4**, 6155 (1992)
21. P. Norby *et al.*, Journal of Solid State Chemistry, **119**, 191 (1995)
22. C.R. Natoli *et al.*, Eur. Phys. J. B **4**, 1 (1988); Z.Y. Wu *et al.*, Phys. Rev. B **55**, 2570 (1997)
23. M. Fabrizio *et al.*, Phys. Rev. Lett. **80**, 3400 (1998);
24. B.J. Sternlieb *et al.*, Phys. Rev. Lett. **76**, 2169 (1996);

25. P.G. Radaelli *et al.*, Phys. Rev. B **55**, 3015 (1997-I);
26. T. Huhne and H. Ebert, Solid St. Com. **109**, 577 (1999);
27. T. Huhne *et al.*, Phys. Rev. B **58**, 10236 (1998);
28. J. Garcia *et al.*, in the present revew;
29. D. H. Templeton and L. K. Templeton, Phys. Rev. B **49**, 14850 (1994-I);
30. I. S. Elfimov *et al.*, Phys. Rev. Lett. **82**, 4264 (1999);
31. M. Takahashi *et al.*, J. of the Phys. Soc. of Japan **68**, 2530 (1999);

Orbital Ordering and metal-insulator transition in V_2O_3

Massimiliano Cuozzo[1,3], Yves Joly[2], El Kebir Hlil[2] and Calogero R. Natoli[1,2]

[1] *Laboratori Nazionali di Frascati, INFN, Casella Postale 13, I-00044 Frascati*
[2] *Laboratoire de Cristallographie, CNRS, associé à l'Université Joseph Fourier, B.P. 166, F-38042 Grenoble Cedex 9*
[3] *Universitá degli Studi Federico II di Napoli, Dpt Scienze Fisiche, Via Cinzia, 80125 Napoli - XII ciclo di dottorato*

Abstract. We reexamine the ground state properties of the Antiferromagnetic Insulating Phase of V_2O_3 in the light of the recent experimental observations by L. Paolasini et al [1] based on Resonant X-ray Scattering at the V K-edge and non-resonant magnetic scattering. We propose a model capable of reconciling their finding that the V atoms are in an S = 1 state with the concomitant evidence of an Orbital Ordering of the magnetic electrons. Realistic calculations for the response function of the non-magnetic resonant scattering based on the $3d$ charge distribution suggested by the model show an amazingly good agreement with the energy and azimuthal scans presented in ref. [1] Its relevance to the physics of the low-temperature (150 K) metal-insulator transition is also briefly discussed in the light of the inelastic neutron scattering data.

INTRODUCTION

This paper aims at providing a consistent picture of the Antiferromagnetic Insulating Phase (AFI) of V_2O_3, capable of reconciling the different pieces of experimental information that are emerging on the basis of neutron and light scattering experiments [1-3]. As is well known, V_2O_3 is considered to be the prototype of the Mott-Hubbard systems, showing metal-insulator transitions from the paramagnetic metallic (PM) phase to the AFI phase at low temperatures ($\approx 150K$) and from the PM phase to a paramagnetic insulating (PI) phase at higher temperatures ($\approx 600K$), due to the interplay between band formation and electron Coulomb correlation. Actually this material is the only known example among transition-metal oxides to show a PM to PI transition. No doubt, interest in this material (and in general in Mott systems) has been rekindled by the unexpected discovery of high T_c cuprate superconductors. To be precise, the cuprates are classified as charge transfer systems, according to the classification by Zaanen et al [4], but similar low-energy physiscs is expected in both classes of materials. Moreover, the direct finding of static orbital-spin correlations in the AFI phase in the case of X-ray scattering [1] and the indirect, but equally important, evidence [2,3] of dynamic orbital-spin correlations in both the PM and PI phases in the case of neutron scattering, make the analysis of the V_2O_3 system rather tempting in view of the key role played by orbital ordering (OO), degeneracy and orbital-spin correlations in the physics of manganites. Indeed fruitful cross-fertilization might result from a deeper understanding of the various ingredients determining the behavior of V_2O_3.

DRAWBACKS WITH THE EXISTING MODELS

At first sight, the observations by Paolasini et al [1], showing evidence for the existence of an orbital ordering with the periodicity predicted by Castellani et al [5], seem to provide a nice confirmation of their model. However this latter was based on the assumption, reasonable at the time in the lack of further experimental evidence, that there was only one magnetic electron in the doubly degenerate e_g band, and that therefore each Vanadium atom was substantially in a state of spin S=1/2. There is now definite evidence in the same ref. [1], based on the non resonant magnetic scattering, that the ratio $<L>/(2<S>) = -0.3$ so that the spin moment is $2<S> = 1.7$, compatible with a spin $S = 1$ state of the Vanadium atoms. This is a strong indication that in the AFI phase of V_2O_3 intraatomic correlations prevail over band delocalization, contrary to the assumption made in ref. [5].

Actually this situation was examined in the second paper of the series [6], where a realistic calculation was performed in the Hartree-Fock (HF) approximation, using a bare tight binding (TB) band structure for the a_{1g} and e_g orbitals in the Hubbard Hamiltonian. This method is very akin to the more modern LDA + U method [8]. There it was shown that for $J/U > 0.2$, where J and U are the intraatomic exchange and Coulomb parameter respectively, a stable self-consistent solution was found with the real (ie observed) spin structure, no orbital ordering, 1.5 electrons in the e_g and 0.5 in the a_{1g} bands aligned by intraatomic exchange, with a spin moment of $1.7\mu_B$. This solution was stabilized with respect to all others by the monoclinic distorsion, with the concomitant opening of a gap for convenient, reasonable values of U and J. Substantially the same solution with the same stabilization mechanism has been very recently put forward by Ezhov et al [9], with no surprise since the two methods are conceptually identical and the range of parameters and the initial bare density of states rather similar. At the time this solution was discarded in favour of solutions with $J/U < 0.2$ showing an orbital ordering, for various reasons, some of which illustrated in ref. [6]. In the light of the new findings we are compelled to reconsider it and examine whether it stands the test of the experimental evidence.

As it is, the solution with $J/U > 0.2$ is "unnatural". For a value of $J/U \approx 0.3$, for example, the authors in ref. [9] find an occupation of 0.95 electrons in each e_g subband. Therefore the ferromagnetic (F) coupling along the vertical pair and in the basal plane along the bond contained in the plane perpendicular to the monoclinic $\vec{b}$ axis should be very weak, since it corresponds to the case of a half filled band, which notoriously favours antiferromagnetic (AF) coupling. In fact the ground state found in ref. [9] strongly competes with the simple AF state, which is lower in energy if the lattice has the symmetry of the corundum phase and is higher by only 80 K (≈ 7 meV/per V atom) in presence of the monoclinic distorsion. It is clear that the real AF state is stabilized with respect to the simple one by the magnetic interactions with the nine next nearest neighbors, which in the real spin structure show six V atoms with spin antiparallel to a given atom in the lattice and only three with spin parallel. The reverse is true for the simple AF state. The consequences of this situation are evident in the calculated values [9] for the exchange integrals, which with reference to Fig. 1 of ref. [9] turn out to be: $J_\gamma = J_\alpha = 4.2\ meV, J_{\beta_1} = -19\ meV, J_{\beta_2} = -8\ meV$, to be compared with the experimental values found by Word et al [10]: $J_\gamma = 27\ meV, J_\alpha = 23\ meV, J_{\beta_1} = J_{\beta_2} = -46\ meV, J_\delta = -2.8\ meV, J_\epsilon = 4.8\ meV$, where a positive J implies a F coupling. Note that these last values are reasonably consistent within the errors with the relation $J_{\beta_1} + J_\delta + 2J_\epsilon = -33(2)\ meV$, found in ref. [3]. We have converted the different sign conventions in the various papers to the one used in ref. [9]. Clearly a stabilizing mechanism at the level of nearest neighbors is missing in the solution found in ref. [9].

Another drawback of this solution is that it fails to reproduce the resonant non magnetic signal observed in ref. [1] at the forbidden (111) reflection. In fact, as shown below, this signal depends on a linear combination with structure phase factors of the difference of the anomalous scattering factors between atoms 4 and 8, 6 and 5, 2 and 3, 1 and 7, using the numbering of fig. 3 in ref. [5]. Consequently, the mechanism proposed in ref. [9] to explain the signal is ineffective, since it invokes a different $|e_g> \pm \alpha |a_{1g}>$ admixture for neighboring basal planes with distance $c_H/6$, in each plane the admixture having a constant sign. However all the above pairs either belong to atoms lying in planes with distance $c_H/3$ (ie they occupy every other plane along the direction of the exagonal c_H axis) or lie on the same plane. Hence their form factor is zero.

OUT OF THE IMPASSE

In order to get out of the *impasse* and to produce a model consistent with all the relevant experimental findings (ie spin $S = 1$ on each V atom, the correct order of magnitude for the exchange integrals, the existence of a resonant scattering signal at the (111) reflection and obviously the spin structure) we observe that out of the three one electron states $|e_g^+>, |e_g^->$ and $|a_{1g}>$ one can form three two electron states: $|e_g^+ e_g^-> \equiv |0>, |a_{1g} e_g^+> \equiv |+>, |a_{1g} e_g^-> \equiv |->$, which even in the presence of the small trigonal distorsion Δ_t in the corundum phase are nearly degenerate (they constitute the spin and orbital triplet ground state of a two electron system in a strong cubic crystal field [11]). Actually there seems to be experimental evidence (no inisotropy in Knight shift and NMR measurements) that in all phases of V_2O_3 the trigonal field splitting Δ_t is nearly zero [12]. However we believe that such experimental evidence is compatible with the existence of a $\Delta_t > 0$ (of the order of 50 meV), due to the mixing of all three states brought about by the band delocalization mechanism. In this field, the invariant state $|0>$ would lie lowest, followed by the doublet $|+>, |->$.

Out of these three states, we can then construct by linear combination three new states with coefficients depending on the amplitude α_0 of the state $|0>$, namely:

$$\begin{aligned}|\Phi_1> &= \alpha_0|0> + \alpha_+|+> - \alpha_-|-> \\ |\Phi_2> &= \alpha_0|0> - \alpha_-|+> + \alpha_+|-> \\ |\Phi_3> &= \sqrt{1-2\alpha_0^2}\,|0> - \alpha_0|+> - \alpha_0|->\end{aligned} \quad (1)$$

where $\alpha_\pm^2 = (1-\alpha_0^2 \pm \sqrt{1-2\alpha_0^2})/2$.

This solution with real coefficients is possible only for $\alpha_0^2 \leq 1/2$. In states $|\Phi_i> (i=1,2)$ the ratio κ of a_{1g} to e_g occupancy per orbital type is given by $\kappa = 2(1-\alpha_0^2)/(1+\alpha_0^2)$ and is obviously linked to the same ratio in the energy region of the bare density of states under the Fermi level. With reference to Fig. 1 of ref. [9], we see that in the monoclinic phase this ratio is given by $0.5/0.75 \approx 2/3$, so that, taking this value for κ, $\alpha_0^2 = (2-\kappa)/(2+\kappa) \approx 1/2$. For the corundum phase the same ratio is ≈ 0.6, so that $\alpha_0^2 \approx 0.55$ [6].

We now make the *ansatz* that κ is slightly greater/less than $2/3$ in the monoclinic/corundum phase (α_0^2 less/greater than $1/2$), based on the fact that the octahedral distortion in the AFI acts in such a way as to localize the e_g orbitals along the directions of the ferromagnetic spin coupling, whereas it affects the hopping a_{1g} orbitals only along one bond in the basal planes, the one along the vertical pairs remaining substantially unaffected. This is because the effective hopping integral is the sum of a direct $3d-3d$ contribution and a covalent contribution, which tend to cancel in all cases mentioned above, except for a_{1g} orbitals along the exagonal c_H axis where they are always in phase (see Eq.s (2.16), (2.17) and Table I in ref. [6]). In other words, due to the monoclinic distorsion the value of α_0^2 switches from a value greater than $1/2$ in the corundum PM phase, where the solution in Eq. 1 does not exist, to a value less than $1/2$ in the AFI phase, where that solution exists.

We then see that in the AFI phase the population of e_g orbitals is asymmetric in the two states $|\Phi_i> (i=1,2)$, the asymmetry parameter being $\Delta_o = \alpha_+^2 - \alpha_-^2 = \sqrt{1-2\alpha_0^2}$ and one can take advantage of this freedom to lower the variational total energy of the system. These states lie lower in energy compared to the third one $|\Phi_3>$ even if $\Delta_t = 0$, since this latter does not have the additional variational flexibility of the first two when the kinetic energy term in the hamiltonian is turned on.

Indeed, following now the same arguments used in ref. [5] for the $|a_{1g}e_g^\pm>$ part of the states $|\Phi_i> (i=1,2)$, we see that, given the real spin ordering, we increase the jumping possibilities (and therefore lower the kinetic energy) if we populate all Vanadium sites according to the orbital occupation schemes found in ref. [5], replacing this time the $|e_g^\pm>$ orbitals with the states $|\Phi_i>$. In fact, for $\alpha_0^2 \to 0$ these states reduce to the eg orbitals used in ref. [5], with the a_{1g} orbital acting as spectator. In particular we find as before two competing wavefunction ordering, the simple AO ordering based on the $|e_g^i>$ (i=1,2) orbitals and the RO(2) ordering based on the basis $|e_g^\pm>= (|e_g^1> \pm |e_g^2>)/\sqrt{2}$ [5].

This situation, depicted in the atomic limit, is born out by preliminary variational calculations based on the crystal wavefunction:

$$|\Psi_{xtal}>= [\Pi_{\vec{\rho}_i\sigma}(\sum_{m\neq m'}^{1,3} A_{\vec{\rho}_i\sigma}^{mm'} \sum_{\vec{k}} G_{\vec{\rho}_i\sigma}^{mm'}(\vec{k}) c^\dagger_{\vec{k}m\vec{\rho}_i\sigma} c^\dagger_{-\vec{k}m'\vec{\rho}_i\sigma})]^{N_c}|0> \qquad (2)$$

Here $|0>$ is the vacuum state, N_c is the number of unit monoclinic cells, $c^\dagger_{\vec{k}m\vec{\rho}_i\sigma}$ is the Fourier transform (FT) of the creation operator for an electron in a Wannier state centered at site $\vec{R}_n + \vec{\rho}_i$ ($\vec{\rho}_i$ denoting the position of atom $i = 1, 8$ inside unit cell n), with orbital type $m = 1, 2$ for orbitals $|e_g^\pm>$ or $|e_g^i>$ and $m = 3$ for orbital $|a_{1g}>$, and with spin σ, as described in ref. [6], Eq. (3.6). Finally $G_{\vec{\rho}_i\sigma}^{mm'}(\vec{k})$ is the FT of the wavefunction in configuration space between two $3d$ electrons with correlation lenght of the order of the interatomic distances.

This kind of wavefunction for the entire crystal extends to a periodic situation the type of correlated *atomic* wavefunction given in Eq. 1, so that we should now think of the states $|\Phi_i>$ as long-lived resonances whose type is correlated throughout the crystal, a situation which is at the heart of the present electron correlation problem.

Its HF approximation is easily depicted as showing an ordered occupation pattern per site of $\alpha_0^2 + \alpha_\pm^2$ electrons of type $e_g^\pm$ and $1 - \alpha_0^2$ electrons of type a_{1g}, with an order parameter $\Delta_o = \sqrt{1 - 2\alpha_0^2}$ describing the unbalanced occupation of the e_g orbitals on different sites.

This situation is at variance with a uniform occupation of $1 + \alpha_0^2$ electrons of e_g type (and $1 - \alpha_0^2$ of a_{1g}) found in ref. [9,6]. We speculate that this new orbitally ordered HF state should be lower in energy than the uniform one and thus provide the missing stabilizing mechanism for the real spin structure and the correct order of magnitude for the exchange integrals.

CALCULATION OF THE ATOMIC SCATTERING FACTOR: COMPARISON WITH THE EXPERIMENTS

Leaving aside the detailed examination of this HF state (and its correlated counterpart in Eq. 2) for future investigation, what matters now is to check that the consequences of the model are in keeping with the non magnetic resonant scattering data of ref. [1].

As is well known the cell Anomalous Scattering Factor (ASF) is given by $F = \sum_i^{1,8} e^{i\vec{q}\cdot\vec{\rho}_i} f_i(\omega)$ where $f_i(\omega)$ is the ASF of the Vanadium atom at position $\vec{\rho}_i$ in the monoclinic unit cell of V_2O_3. Following the notation by Dernier and Marezio [14] and the numbering of V atoms in the monoclinic unit cell as in ref. [5], we find for F at the general reflection (h, k, l):

$$F = (bf_4 + sf_8) + (bf_5 + sf_6)e^{-4i\pi(x+y+z)} + (f_2 + bsf_3)e^{-4i\pi y}e^{i\pi(k+l)} \quad (3)$$
$$+ (f_1 + bsf_7)e^{-4i\pi(x+z)}e^{i\pi(k+l)}$$

where $x = 0.3438, y = 0.0008, z = 0.2991$ [14], $b = e^{i\pi(h+k+l)}$ and $s = -1$ if magnetic reflections are considered, since sites 8, 6, 3 ,7 have opposite spin orientation than sites 4, 5, 2, 1. Otherwise s = 1.

At the (1,1,1) reflection we see that the magnetic scattering is zero, whereas for non magnetic scattering we obtain

$$F = (f_8 - f_4) + (f_6 - f_5)e^{-4i\pi(x+y+z)} + (f_2 - f_3)e^{-4i\pi y} + (f_1 - f_7)e^{-4i\pi(x+z)} \quad (4)$$

This can also be written, using the symmetry operations of the monoclinic space group $I2/a$ with the same notation as in ref. [5] (pag. 4954),

$$F = (1 + e^{-4i\pi(x+y+z)}I - e^{-4i\pi y}C - e^{-4i\pi(x+z)}C_2)(f_8 - f_4) \quad (5)$$

implying that $F = 0$ if $f_8 - f_4 = 0$.

It is important to note that the monoclinic unit cell used in ref. [14] is not primitive with respect to the lattice (being body-centered its volume is double the primitive cell) and that sites 8, 6, 2, 1 with their oxygen environment are translationally equivalent to sites 4, 5, 3, 7. Therefore the observation of a non magnetic resonant signal at the (1,1,1) reflection cannot be due to a coherent distorsion of the invironment of sites 8 and 4, as in $LaMnO_3$ [13], due to their equivalence. Consequently, to account for this signal, one must assume a coherent organization of the crystal, whether of structural or electronic origin, that doubles the primitive cell and makes the monoclinic unit cell of ref. [14] primitive. The structural origin of the effect is ruled out by the fact that this latter cell already includes the description of the translational properties of the crystal, even though it is well possible that within the experimental errors a coherent distorsion might still be present, giving the observed signal. In this latter case however one would expect a much bigger signal in the energy region of the rising edge (*ie* between 5.47 and 5,48 KeV), besides the one detected at 5.464 KeV in correspondence to the $3d - T_{2g}$ transition, since the $4p$ conduction band states are much more sensitive to atomic distorsions. But no signal is present in correspondence of the rising edge (Fig. (2) of ref. [1]). A signal of electronic origin might arise either from an ordered distribution of the quadrupolar moments related to different admixtures of e_g and a_{1g} orbitals on the Vanadium atoms, as suggested in ref. [9], or from an ordered occupancy of the e_g orbitals with order parameter Δ_o as suggested in the present paper. In any case, in order to obtain a signal at the (1,1,1) reflection, it is mandatory that the new periodicity be such that sites 8 and 4, respectively at the corner and the center of the monoclinic unit cell, are inequivalent. This condition immediately rules out the quadrupole moment ordering suggested in ref. [9] and the simple AO orbital ordering found in ref. [5], but is compatible with the RO(2) type of ordering, as shown in ref. [1]. Note that this latter is the only ordering that is left invariant by the application of the generators of the space group $I2/a$ of the monoclinic phase (see Eq. 5 above).

In order to calculate the ASF $f_i(\omega)$ of the single Vanadium atom at site j we follow the one-electron cluster method used in ref. [13], whereby

$$f_j(\omega) = (\hbar\omega)^2 \int_{E_f}^{\infty} dE \sum_L \frac{<1s|T_s^\dagger|\Psi_L(E)><\Psi_L(E)|T_i|1s>}{\hbar\omega + E_{1s} - (E + I_{1s}) + i\Gamma(E)} \qquad (6)$$

As in ref. [13], $\hbar\omega$ is the photon energy, I_{1s} is the ionization energy of the $1s$ level up to the Fermi level E_f and $|\Psi_L(E)>$ is the continuum scattering solution of the Dyson (Schrödinger-like) for the cluster of atoms around the central V atom at energy E above E_f in response to an exciting wave $J_L(\vec{r})$ calculated using the real part of the Hedin-Lundquist (HL) self-energy. $\Gamma(E)$ is then the sum of the imaginary part of this latter plus the half-width at half-maximum of the core hole state. Moreover for the transition operator T_i we take $\vec{\epsilon}_i \cdot \vec{r}(1 + i\vec{k}_i \cdot \vec{r})$, where $\vec{\epsilon}_i$ and $\vec{k}_i$ are respectively the polarization and propagation vector of the incident light, since the V atom is not in a centrosymmetric position relative to its surrounding. The same holds for the transition operator T_s of the scattered light. In the $\sigma - \sigma$ channel and for forward scattering $T_i \equiv T_s$

In the lack of a self-consistent charge density with OO we use an input charge density obtained by superposition of atomic charge densities. In particular for Vanadium atoms we take the configuration $3d^3 4s^2$. However we spherically average the charge density of one of the three $3d$ electrons which, together with the two $4s$ electrons, is supposed to fill the oxygen $2p$ states to create the covalent bond. Finally the charge density of the remaining two $3d$ electrons is derived from the correlated wavefunctions $|\Phi_i> (i = 1,2)$ distributed among the various sites according to the RO(2) OO (ie we populate each site with $1 - \alpha_0^2$ electrons of type a_{1g} and $\alpha_0^2 + \alpha_\pm^2$ electrons of type $e_g^\pm$ according to whether the Φ function is of type 1 or 2). From this charge density the Coulomb potential and the HL self-energy is easily constructed. The resulting Dyson equation with scattering wave boundary conditions for the considered cluster of atoms is then solved by the Finite Difference Method (FDM), which discretizes the Laplacian and makes no approximation on the shape of the optical potential (see ref. [13] and ref.s [9,10] therein). In this context, also the Poisson Equation is solved by the same method, with the input charge density generated in the way described above. Once the $|\Psi_L(E)>$ states are so obtained, the expressions for the atomic form factor in Eq. 6 and correspondently the cell form factor are calculated. As a by-product the imaginary part of the $\sigma - \sigma$ forward channel provides the absorption coefficient, which is extremely helpful in the interpretation of the anomalous scattering experiments.

Due to the large memory requirement of the FDM program in the case of low symmetry, as the present compound, we limited ourselves to a cluster of eleven

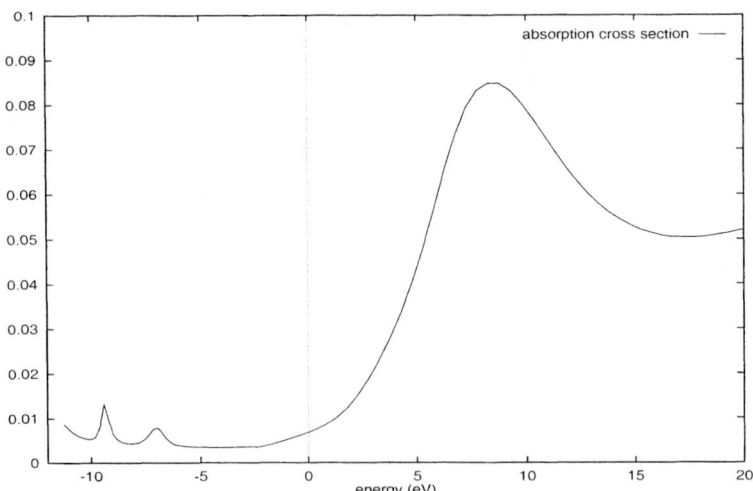

FIGURE 1. Vanadium K-edge absorption cross section in Mbarns for the eleven atom cluster VO_6V_4. The two prepeaks are due to dipolar and quadrupolar transitions to $3d$ states of T_{2g} and E_g character in the approximately octahedral oxygen environment

atoms (the central Vanadium, the six oxygen nearest neighbors and the four Vanadium nearest neighbors: VO_6V_4). Since we are probing the $3d$ states, which are fairly well localized, we expect this cluster to be sufficient for a realistic representation of the local electronic properties of V_2O_3. We need only perform two calculations, one centered on site 8 and the other on atom 4. The remaining six on the other sites are generated by application of the symmetry operations I, C_2, C, as shown in the expression of Eq. 5. The modulus squared of this latter is then proportional to the observed intensity.

Fig. 1 shows the Vanadium K-edge absorption coefficient for this cluster. Apart from details which are reproduced by increasing the size of the cluster, the general shape is very much similar to the measured one [1]. In particular the two pre-edge peaks represent transitions to $3d$ states of T_{2g} and E_g character, if one neglects in first approximation the distortion of the oxygen octahedra around the V atoms in the monoclinic AFI phase. Indeed, due to these distortions, their strenght is partly due to dipolar transitions which account for half of the observed intensity, the other half being due to the quadrupole operator. Therefore both mechanisms must be considered for a realistic calculation of the resonant signal. The first peak contains six states, of which only two are occupied, so that the Fermi Level lies somewhere before the maximum.

Fig. 2 shows an energy scan of the resonant signal in the region around the two pre-peaks for the $\sigma - \sigma$ and $\sigma - \pi$ channels, calculated using the formula in Eq. 6 with $\Gamma(E) = 0.5\,eV$, roughly equal to the half-width of the Vanadium $1s$ core hole state, since the imaginary part of the HL self-energy is nearly zero at these excitation energies. We have neglected the experimental energy resolution. The

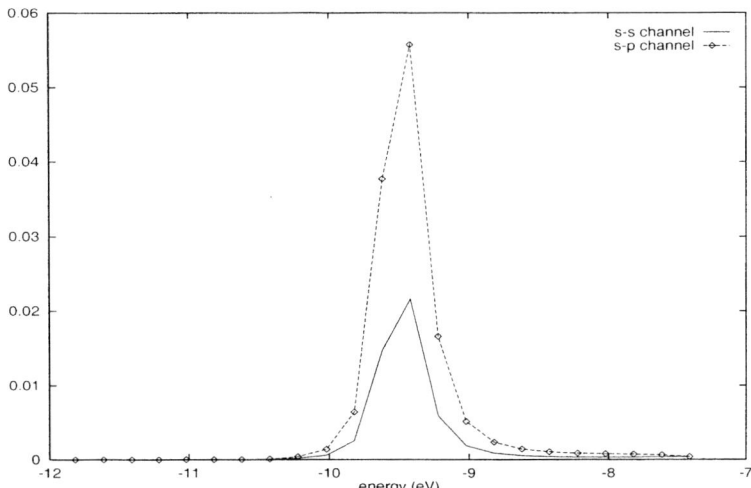

FIGURE 2. Energy scan of the resonant scattering signal at the V K-edge in the region around the two pre-peaks of Fig. 1 for the $\sigma - \sigma$ and $\sigma - \pi$ channels

most important features of the experimental findings are here reproduced. Indeed the ratio of the $\sigma - \pi$ channel to the $\sigma - \sigma$ channel intensity is 2.7 to 1, very close to the measured value 2.2 to 1 and there is no intensity in the energy region of the E_g states, in keeping with the fact that only the empty T_{2g} states show an orbital order in response to a similar order of the occupied ones. Also no signal is obtain in the energy region of the $4p$ conduction band rising edge, some 15 eV above (not shown). Finally the measured width of the resonant signal (1.6 eV) is clearly resolution limited. Note that the intensity of the two channels is proportional to the OO parameter Δ_o^2, since the amplitude is proportional to Δ_o. The present calculations correspond to a value of $\Delta_o = 0.3$. However, in the lack of an absolute calibration, it is impossible to fit Δ_o on the experimental data.

Fig.s 3 and 4 show respectively the azimuthal scans around the momentum transfer vector for the $\sigma - \sigma$ and $\sigma - \pi$ channels at an energy corresponding to the maximum of the signal in the energy scan (-9.4 eV), together with the experimental points graciously provided by L. Paolasini. We observe that the agreement with measurements is amazingly good in the first case, since the peaks at 90° and −90° are of equal intensity due to the large error bars of the peak points at −90° (cfr Fig. (3) of ref. [1]). Less good is the agreement in the $\sigma - \pi$ channel, though still very reasonable. Note also that these scans are extremely sensitive to the energy value at which they are recorded, as we have checked in our calculations.

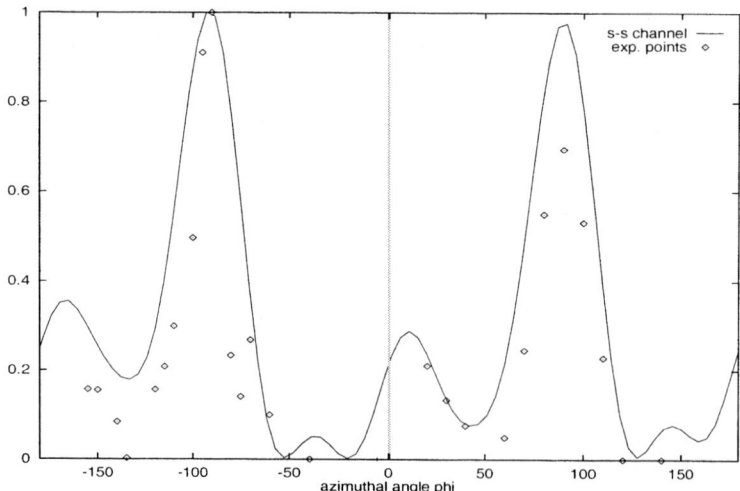

FIGURE 3. Azimuthal scan of the resonant scattering signal in the $\sigma - \sigma$ channel at the energy corresponding to the maximum of the signal in Fig. 2 (-9.4 eV)

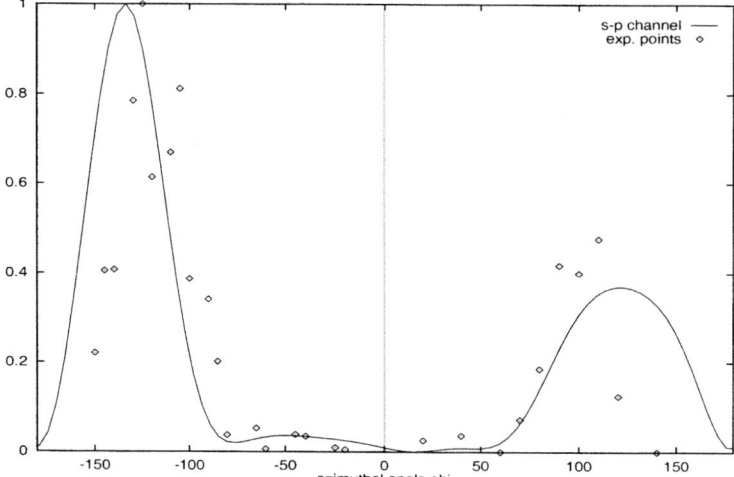

FIGURE 4. Azimuthal scan of the resonant scattering signal in the $\sigma - \pi$ channel at the energy corresponding to the maximum of the signal in Fig. 2 (-9.4 eV)

CONCLUSIONS

Summarizing, the assumption that the two antibonding $3d$ electrons of the V^{3+} ion in a trigonal field with $\Delta_t \geq 0$ are coupled to spin $S = 1$ in states $|\Phi_i\rangle$ $(i = 1, 2)$ as given by Eq. 1 and distributed among the various sites according to the RO(2) Orbital Ordering, leads to a charge density and to a self-energy of the HL type such that, when inserted in the corresponding Dyson equation to calculate the intermediate excited states in the expression for the Anomalous Scattering Factor Eq. 5, give a resonant signal in very good agreement with measured quantities like the energy and the azimuthal scans for both the $\sigma - \sigma$ and $\sigma - \pi$ channels. A more detailed analysis using the variational wavefunction in Eq. 2 will be given elsewhere. However at this stage one can already draw some interesting conclusions regarding the physics of the metal insulator transition in V_2O_3. In fact the inelastic neutron scattering experiments by Wei Bao et al [3] on this compound show that the phase transition to the AFI state from the PM and the PI states "introduces a sudden switching of magnetic correlations to a different spatial periodicity which indicate a sudden change in the underlying spin Hamiltonian. To describe this phase transition and also the unusual short-range order in the paramagnetic state, it seems necessary to take into account the orbital degrees of freedom associated with the degenerate d orbitals at the Fermi level in V_2O_3". Moreover "the fact that the magnetic transition is strongly first order indicates that the *intrinsic* Néel temperature, T_N^{AFI}, of the spin Hamiltonian associated with the AFI phase is larger than the orbital ordering temperature T_O". Now in the old model by Castellani et al [5,6], corresponding to the case $\alpha_0 = 0$ in Eq. 1, the excitation energies associated with the orbital ordering is of the same order of magnitude as that associated with the spin ordering, *ie* of the order of $t^2 J/U^2$, where as usual t is the intersite hopping integral. In the present model however, the orbital excitation energies are reduced by a factor $\Delta_o = \sqrt{1 - 2\alpha_0^2}$ with respect to spin excitations energies, exactly in keeping with the mechanism proposed by Wei Bao et al [3].

REFERENCES

1. L. Paolasini et al, *Phys. Rev. Lett.* **82**, 4719 (1999).
2. Wei Bao et al, *Phys. Rev.* **58**, 12727 (1999);
3. Wei Bao et al, *Phys. Rev. Lett.* **78**, 507 (1997)
4. J. Zaanen G.A. Sawatzky and J.W. Allen, *Phys. Rev. Lett.* **55**, 418 (1985)
5. C. Castellani, C.R. Natoli and J. Ranninger, *Phys. Rev.* B **18**, 4945 (1978)
6. C. Castellani, C.R. Natoli and J. Ranninger, *Phys. Rev.* B **18**, 4967 (1978)
7. R.B. Moon, Phys. Rev. Lett. **25**, 527 (1970)
8. V.I. Anisimov, J. Zaanen and O.K. Andersen, *Phys. Rev.* B **44**, 943 (1991); V. Anisimov, F. Aryasetiawan and A. Lichtenstein, *J. Phys. Condens. Matter* **9**, 767 (1997)

9. S.Y. Ezhov, V.I. Anisimov, D.I. Khomskii and G.A. Sawatzky, *Phys. Rev. Lett.* **83**, 4136 (1999)
10. R.E. Word *et al*, *Phys. Rev.* **23**, 3533 (1981)
11. C.J. Ballhausen, *Introduction to Ligand Field Theory*, McGraw-Hill, New York (1962)
12. M. Rubistein, *Phys. Rev.* B **2**, 4731 (1970); see discussion on pag. 4738
13. M. Benfatto, Y. Joly and C.R. Natoli, *Phys. Rev. Lett.* **82**, 636 (1999)
14. P.D. Dernier and M. Marezio, *Phys. Rev.* **2**, 3771 (1970)

Quantum Monte-Carlo calculations and possible impact on angle resolved photoemission spectroscopy

W. Schattke

*Institut für Theoretische Physik und Astrophysik,
Christian-Albrechts-Universität zu Kiel,
Leibnizstrasse 15, D-24098 Kiel*

Abstract. Quantum Monte-Carlo calculations for the ground state are well established, and a few work exists already on excited states. For the purpose of band structure investigations with angle resolved photoemission further development with respect to excitation and transport is still necessary on a Monte-Carlo basis to take full advantage of its many-body capacity and to generalize the one-step model of photoemission. As a surface sensitive technique the method has especially to address to surface systems.

INTRODUCTION

Current density functional theory (DFT) calculations base on local density or more refined functionals which evaluated for the homogenous system bear an uncontrolled approximation if transferred to the inhomogenous case as e.g. to the solid where it has been nevertheless applied with overwhelming success. The validity of the approximations in the actual investigations and their extension is not clear and cannot be quantified. A variational method basing on the minimum energy principle offers a tool for quantitative comparison and for further developing the functional with respect to many-body effects.

The variational Quantum Monte-Carlo (QMC) method minimizes the total energy of a system of about 1000 electrons representing the solid with respect to the parameters in a wave function ansatz. Importance sampling is used to integrate the energy expectation value. The parameters represent one-body quantities in the one-particle wave function of the Slater determinant, many-body properties via the Jastrow factor accounting for the correlation and involve additional geometrical structure parameters. [1]

Reliable results are available for light elements, such as Lithium [2–4], and for a few semiconductors [5–8]. They refer to the bulk case only and have been obtained for Lithium with the all-electron potential and for semiconductors with a pseudopotential. Lattice constant, total energy, bulk modulus and the density distribution are the main observables to be determined. The results compare well with density functional calculations and compete even with the most advanced techniques as e.g. the generalized gradient approximation.

It is possible to extract the exchange correlation functional for the DFT calculations from the QMC results for inhomogenous systems by sampling the two-particle correlation function, thus generalizing the procedure for homogenous systems. [9] First steps towards the simulation of excited states have been carried through regarding the excitation process embedded in an N-particle system and sampling directly differences of total energies via diffusion QMC [10,11], and indirectly, through sampling the one-body density matrix with a many-body Slater-Jastrow ansatz and evaluating their eigenvalues via variational QMC [12].

Current investigations could extend the QMC treatment towards solid surfaces showing that the relaxational energies of the GaAs(110) surface are statistically resolvable [14]. The main problem to be overcome was to reduce the pressure exerted by the mutual electronic Coulomb repulsion as represented by the Jastrow factor. This effect led to an expulsion of the electrons from the solid slab used to describe the surface system. The introduction of counter terms into the wave functions yielded through the confinement to the region of the ions a satisfying optimization of the slab's total energy.

Beyond the static equilibrium investigations the challenge arises in modelling transport properties of electrons. From our point of view the scattering states, as they are indispensable in electron diffraction and photoemission, deserve to be tackled by these methods. It is especially the case of low kinetic energies, say e.g. around and below 100 eV, we are interested in, because of the strong multiscattering processes ocurring in the path of the propagating electrons through the scattering medium of many other particles. Usually, they are determined as wave functions in a one-particle picture by solving the Schrödinger equation. They show strong structures in direct geometrical space depending sensitively on the optical potential, illustrating the importance for calculating the scattering t-matrix and the matrix element of photoexcitation on a many-body basis. Such a QMC scheme should yield not only the self-energy, the optical potential being its imaginary part, but include also the possibility of direct non mean field individual losses during scattering or excitation.

In the next section a short sketch of one-step calculations is given. The subsequent section deals with the description of QMC calculations for the ground state of bulk and surface systems, where special attention is paid to the rather new developments for surfaces. The paper ends with the consideration of QMC transport for scattering states and an outlook for a full QMC photoemission theory.

PHOTOEMISSION

The one-step calculation of photoemission is based on the well known golden-rule formula for the intensity I of emitted electrons, viz.

$$I \propto \sum_{ini} |<\Phi^{\star}_{LEED}(E_{fin},\mathbf{k}_{\|})|(\mathbf{A}\cdot\mathbf{p}+\mathbf{p}\cdot\mathbf{A})|\psi_{ini}>|^2 \delta(E_{fin}-E_{ini}-\hbar\omega), \quad (1)$$

detected at final energy and momentum, E_{fin} and $\mathbf{k}_{\|}$, in a state usually represented as a time inverse state of low energy electron diffraction (LEED). The initial state ψ_{ini} together with the δ-function of energy conservation is often written as

$$I \propto \sum_{nm} <\Phi^{\star}_{LEED}(E_{fin},\mathbf{k}_{\|})|(\mathbf{A}\cdot\mathbf{p}+\mathbf{p}\cdot\mathbf{A})|n>$$
$$\mathrm{Im} G_{nm} <m|(\mathbf{A}\cdot\mathbf{p}+\mathbf{p}\cdot\mathbf{A})|\Phi_{LEED}(E_{fin},\mathbf{k}_{\|})> \quad (2)$$

to stress the relation to the spectral density $\mathrm{Im}G$ projected onto a set of states n. The determination of I involves the calculation of the bandstucture for the bound states, a procedure for the LEED states, which arise as scattering states with outgoing boundary conditions, and the integration of the matrix elements. In the subsequent examples, the Green function represented in localized orbitals was fitted to the solution of a self-consistent bandstructure for the initial states [16,17], the Schrödinger equation was solved by matching solutions of the complex bandstructure (presently, the Laue representation with algebraic methods and an ab-initio pseudopotential is used) of a local pseudopotential for the final states [18], and a real space integration was performed to obtain the matrix elements.

There exist simple methods for interpreting the experimental spectra, namely band mapping methods. Their accuracy, however, is restricted to about 200 meV for simple semiconductor surfaces. [18] A kind of short cut such as an inverse photoemission procedure is principally impossible and simple approximate but more accurate techniques than band mapping are still lacking. Thus, to exploit the full information contained in up to date photoelectron spectroscopy, which resides far below the mentioned 200meV resolution, one has to rely on a trial and error procedure, not unfamiliar to other spectroscopies and rather elaborate in the case of photoemission. Even an "ab-initio" claimed bandstructure bears adjustable parameters, on one hand, and is most unsafe about the role of many-body effects, on the other hand, generally demanding for experiment, and an adequate interpretation.

As an example of recent interest, figure 1 shows the theoretical photocurrent from the GaN(0001) surface compared to the experimental electron distribution curve. In theory the energy positions of adlayer emissions can be clearly identified, because of being traced back to certain orbitals in the calculation as demonstrated by the plotted density of states (DOS). Together with the main peak they represent a safe indicator and may thus serve for fixing the absolute energy scale. The plot drastically proves the difficulties in properly adjusting the energy zero of a spectrum, independent of the specific procedure, be it the rather less justified extrapolation by the tangent or even the deconvolution into separate lines whatever

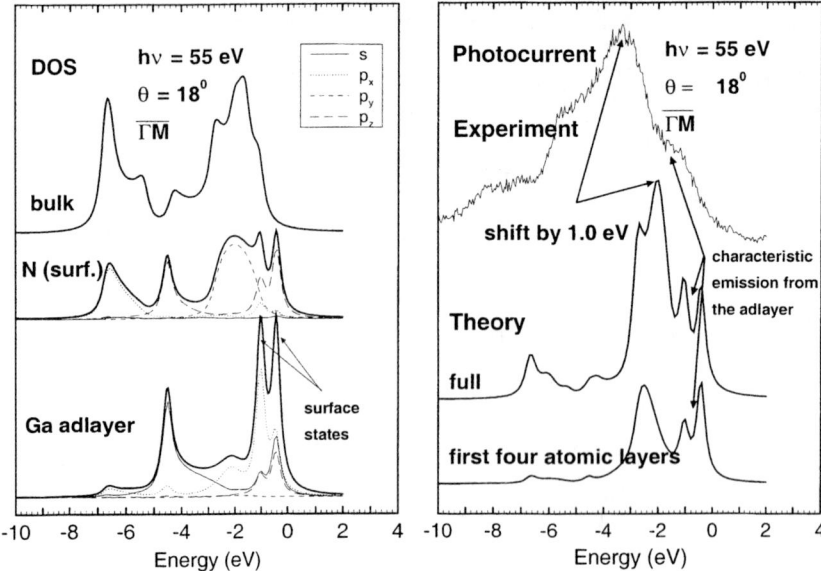

FIGURE 1. Photocurrent (right) and DOS (left) for GaN(0001) with same vertical scale in theoretical PE curves and separately in DOS curves; total DOS decomposed into atomic and orbital contribution; experimental photocurrent positioned as originally, with kind permission of the authors [21].

type of lineshape is chosen. The knowledge of the zero is important for the correct determination of the bandwidth which cannot be definitely decided by experiment alone.

From the difference between the intensities of the full spectrum and that of the first four layers, the origin of the main maximum can be attributed to bulk transitions, which also is supported by the DOS. The spectrum is further analysed with the help of plotting the matrix elements associated with different orbitals, see figure 2. For example, the maximum at -2.5 eV in the contribution from the first four layers carries also emissions from N p_y states (see DOS) supported by a large p_y matrix element. The adlayer emissions are attributed to N p_z for both parts of the twin peak, together with minor contributions from Ga p_x because of its small matrix element, and to Ga s, p_z for the high energy part of this peak. This interpretation is consistent with an analysis of full spectral series. [20]

There is still a vivid interest in valence band photoemission, and the one-step calculations further evolve to grasp the full information supplied by experiment, as e.g. the evaluation for the entire two-dimensional acceptance cone citeprlsolter. Experimental developments as two electron coincidence measurements [23] and time resolved two photon techniques [24] ask for further generalization of the one-step model, on one hand. On the other hand, the acquired high experimental resolution

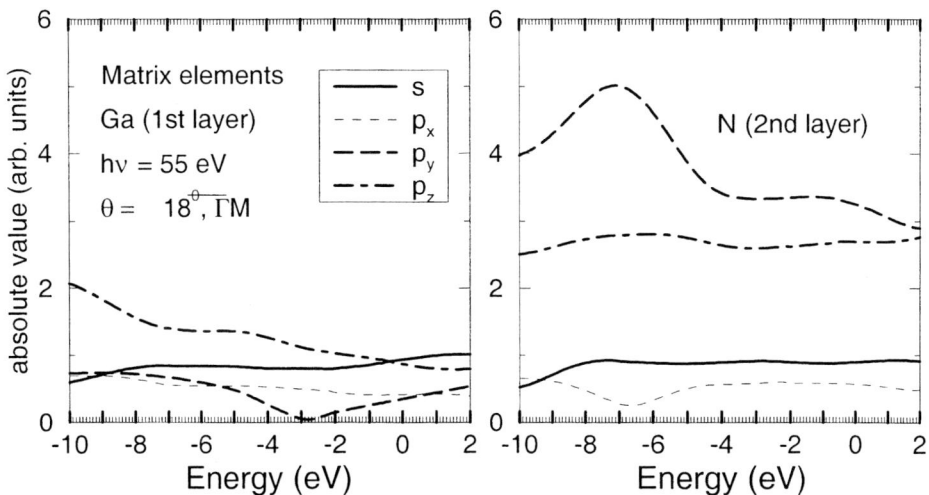

FIGURE 2. Matrix elements for phototransition from various orbitals of first layer Ga (left) and second layer N (right) with vertical scale identical in all plots.

represents a strong demand for an accurate treatment of the many-body effects, as it seems to be presently feasible with QMC methods.

QUANTUM MONTE-CARLO

A Bulk QMC

The statistical evaluation of the energy expectation value of a many-body system of N particles via QMC samples the $3N$-dimensional configuration space through the weight given by the absolute square of the many-body wave function, Ψ_λ.

$$\langle \hat{H} \rangle = \int \frac{|\Psi_\lambda(\mathbf{Y})|^2}{\int |\Psi_\lambda(\mathbf{Y})|^2 d\mathbf{Y}} \frac{\hat{H}(\mathbf{Y})\Psi_\lambda(\mathbf{Y})}{\Psi_\lambda(\mathbf{Y})} d\mathbf{Y} \quad (3)$$

The Metropolis algorithm [28] is used to guarantee importance sampling during the random walk so that the energy can be estimated as the average of the local energy with purely statistical error

$$\langle \hat{H} \rangle \approx E_\lambda := \frac{1}{M} \sum_{i=1}^{M} \frac{\hat{H}(\mathbf{Y_i})\Psi_\lambda(\mathbf{Y_i})}{\Psi_\lambda(\mathbf{Y_i})}, \quad \sigma(E_\lambda) = \sqrt{\frac{1}{M}\mathrm{Var}(\frac{\hat{H}\Psi_\lambda}{\Psi_\lambda})}. \quad (4)$$

Ritz's variational principle ensures that the minimum E_{λ^*} of E_λ in the parameter space is an upper bound for the true ground state energy E_0.

The wave function has to be prescribed by an ansatz with a set λ of variational parameters. The quality of the wave function is controlled either by the minimum property of the expectation value or by its variance, which should vanish for the optimal choice, i.e. the eigenfunction of the hamiltonian. The former criterion is preferred for sequential adjustment of the parameters of the wave function, whereas the latter yields an estimate of the accuracy obtained. The general form uses a decomposition into a Slater determinant Ψ_S with suitably chosen one-particle functions to describe the one-particle properties and a correlation part such as a Jastrow factor Ψ_J to account for the mutual Coulomb repulsion. The one-particle functions may be taken from DFT calculations or more general as an ansatz with sufficient variational freedom to be adjusted in the course of minimization. From our experience, localized functions reflecting the orbital character behave superior over a more plane-wave like set of functions not only in the case of covalently bound systems such as GaAs but also in the case of metals such as Li. Thus, the wave function is supposed as

$$\Psi(\mathbf{r}_1\sigma_1,\ldots,\mathbf{r}_N\sigma_N) = D^\uparrow(\mathbf{r}_1,\ldots,\mathbf{r}_{\frac{N}{2}}) D^\downarrow(\mathbf{r}_{\frac{N}{2}+1},\ldots,\mathbf{r}_N) \times$$
$$\exp\left(-\sum_{i<j} u(\mathbf{r}_{ij},\sigma_{ij}) - \sum_i v_{cf}(\mathbf{r}_i)\right) \quad (5)$$

with σ_{ij} denoting parallel or antiparallel spins. The repulsive two-body part u is chosen

$$u(\mathbf{r},\sigma) = A\frac{1}{r}\left(1 - e^{-r/F(\sigma)}\right). \quad (6)$$

A is a variational parameter and $F(\sigma)$ is, for given A, fixed by the cusp condition to remove the Coulomb singularity. The configuration space is a supercell as large as compatible with available computer resources. To reduce finite size effects (see also ref. [13]) an identical periodic repetition is applied to all wave function coordinates, i.e. at the moment an electron leaves the supercell space during the multidimensional random walk it will be immediately transferred back to it by a superlattice bftor. Since tabulation and interpolation of spatial functions has proven to be best suited for a fast simulation run, all functions defined on the range of the supercell, i.e. the potentials of the hamiltonian as well as the two-body Jastrow part, have to be mirrored into the full periodic space, which is analytically achieved by appropriate Ewald summations.

Figure 3 shows a typical result for bulk GaAs, displaying the total energy as a function of the lattice constant. The minimum lattice constant, and through the parabolic fit, the bulk modulus satisfy the measured values with an accuracy as in refined DFT calulations, see table A. Defining a step by moving once each of the total 256 electrons, the length of a run comprised about several 10^3 steps and took 6 hours on an α-EV56 (600) CPU yielding one energy value for the actual lattice parameter. A nonlocal pseudopotential was used together with atomic orbitals for the one-particle wave functions in the determinant. The minimization yields a

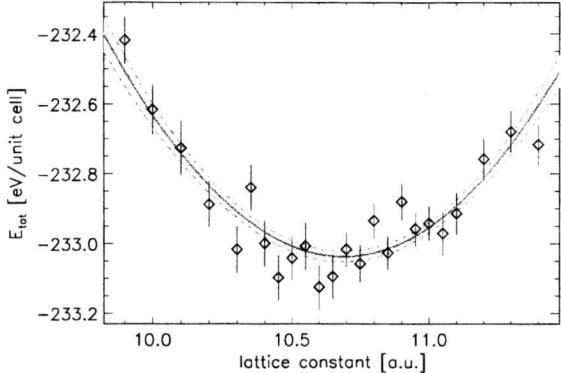

FIGURE 3. Total energy of bulk GaAs vs. cubic lattice parameter in a quadratic parameter fit with displayed width indicating standard deviation and bars denoting statistical error of single QMC runs.

significant compression of the orbitals up to 10% in the solid and 15% in an isolated atom due to the Jastrow factor.

	E_{atom} eV/uc	E_{tot} eV/uc	E_{bind} eV/uc	a_0 Bohr	B kbar
Experiment	−226.773	−233.43	−6.67	10.6831	756
	±0.030	±0.12	±0.09	±0.0001	±4
VQMC LCAO nloc	−228.134	−233.04	−4.90	10.689	786
	±0.008	±0.08	±0.20	±0.080	±100
VQMC LCAO loc	−226.56	−235.10	−8.54	10.13	673
	±0.01	±0.04	±0.05	±0.04	±50
LDA [25]			−8.16	10.41	
LDA [26]			−7.13	10.90	710
GGA [25]			−6.45	10.70	
GGA [26]			−5.76	10.92	680
INC [27]			−6.20	10.67	770

TABLE 1. Comparison of variational QMC (VQMC) calculations with experiment and density functional calculations, such as LDA, generalized gradient approximation (GGA), and an incremental method (INC).

B Surface QMC

It is a nontrivial step to apply the QMC technique to the electronic structure of surfaces. As mentioned above the finite size effects are controlled by periodic repetition of the simulation cell which can be achieved with the help of Ewald summations. This is a significant difference to cluster QMC calculations. In the case of a surface the periodic symmetry along the surface normal has to be broken with the consequence that one has to deal with two-dimensional Ewald sums, with still further enhanced difficulties. The problem arises because of the wish to tabulate the potential exerted on the actually sampled electron by all the other charges in the simulation cell as well as by their repeated images.

A finite slab is used instead of a halfspace system and no periodicity is assumed perpendicular to it. For the long range parts of the potential, i.e. the Coulombic part of the pseudopotential together with the electron-electron interaction, jellium energies of density n_0 are subtracted which cancel each other in the total sum over ions and electrons, just as it occurs also in the three-dimensional case. Denoting by $\mathbf{R}$ the ion and by $\mathbf{r}$ the electron position bftors indexed by the same multiindex $j = \{\alpha, \rho\}$ for the particle ρ in the α image of the simulation cell which itself is a multiple of primitive crystal unit cells, the Coulomb energy E_c is written as

$$2E_c = \sum_i ([-\sum_j \frac{1}{|\mathbf{R}_i - \mathbf{r}_j|} + \int_{\tilde{\Omega}} \frac{n_0 d^3 r'}{|\mathbf{R}_i - \mathbf{r}'|}] + [\sum_j \frac{1}{|\mathbf{r}_i - \mathbf{r}_j|} - \int_{\tilde{\Omega}} \frac{n_0 d^3 r'}{|\mathbf{r}_i - \mathbf{r}'|}] +$$
$$[-\sum_j \frac{1}{|\mathbf{r}_i - \mathbf{R}_j|} + \int_{\tilde{\Omega}} \frac{n_0 d^3 r'}{|\mathbf{r}_i - \mathbf{r}'|}] + [\sum_j \frac{1}{|\mathbf{R}_i - \mathbf{R}_j|} - \int_{\tilde{\Omega}} \frac{n_0 d^3 r'}{|\mathbf{R}_i - \mathbf{r}'|}]) \quad (7)$$

where the multiindex i runs over the simulation cell only, consisting of Ω_0 in parallel direction and the infinite interval perpendicular to it. Zero denominator terms have to be omitted. By $\tilde{\Omega} = (\sum_\alpha \Omega_\alpha) \times [-\frac{z_0}{2}, +\frac{z_0}{2}]$ the decomposition into the infinite surface parallel area times a reduced interval perpendicular to the slab is denoted. The derivation is kept independent of the choice of z_0. Thus, the simulation allows for arbitrary positions of the electrons in the vacuum according to their probability, whereas the jellium is confined to this finite interval. Consider e.g. the first square brackets of equation 7, the j-summation and the parallel integration are broken into single periodic repetitions of the simulation cell denoted by the lattice bftors $\mathbf{S}_\alpha$, viz.

$$\sum_\alpha [-\sum_\rho \frac{1}{|\mathbf{R}_i - \mathbf{S}_\alpha - \mathbf{r}_\rho|} + \int_{\tilde{\Omega}_0} \frac{n_0 d^3 r'}{|\mathbf{R}_i - \mathbf{S}_\alpha - \mathbf{r}'|}]$$
$$= \sum_\rho [\sum_\alpha (-\frac{1}{|\mathbf{R}_i - \mathbf{S}_\alpha - \mathbf{r}_\rho|} + \frac{1}{\tilde{\Omega}_0} \int_{\tilde{\Omega}_0} \frac{d^3 r'}{|\mathbf{R}_i - \mathbf{S}_\alpha - \mathbf{r}_\rho - \mathbf{r}'|}) + \quad (8)$$
$$\sum_\alpha \frac{1}{\tilde{\Omega}_0} \int_{\tilde{\Omega}_{0\rho} \backslash \Omega_0} \frac{d^3 r'}{|\mathbf{R}_i - \mathbf{S}_\alpha - \mathbf{r}_\rho - \mathbf{r}'|}]$$
$$=: -\sum_\rho [v^s(\mathbf{R}_i - \mathbf{r}_\rho) - m_\rho(\mathbf{R}_i - \mathbf{r}_\rho)] \quad (9)$$

with $\tilde{\Omega}_0 = \Omega_0 \times [-\frac{z_0}{2}, +\frac{z_0}{2}]$ and a shift by the bftor $\mathbf{r}_\rho$ has been applied which leaves the parallel integration invariant but shifts the z-integration according to the z-component of $\mathbf{r}_\rho$. The last term in equation 9 yields the error involved, if the shift in the integration interval is suppressed, as it is done within the round brackets of that equation. The advantage of this representation relies upon numerically tabulating the surface potential

$$v^s(\mathbf{R}) = \sum_\alpha \left(\frac{1}{|\mathbf{R} - \mathbf{S}_\alpha|} - \frac{1}{\tilde{\Omega}_0} \int_{\tilde{\Omega}_0} \frac{d^3 r'}{|\mathbf{R} - \mathbf{S}_\alpha - \mathbf{r}'|} \right) \qquad (10)$$

and using a fast interpolation of these data at the actual sites during the Monte-Carlo run. This yields the necessary gain in computer performance.

Lumping the α-summation together the remainder m_ρ in equation 9 consists of an integration over $(\sum_\alpha \Omega_\alpha) \times [-\frac{z_0}{2} - z_\rho, +\frac{z_0}{2} - z_\rho] \setminus [-\frac{z_0}{2}, +\frac{z_0}{2}]$, which may be interpreted as two surface parallel sheets of opposite homogeneous charge distributions and of thickness z_ρ, the negative charge residing on the positive z-half of the slab for $z_\rho > 0$. The z-coordinate of the shift bftor $\mathbf{r}_\rho$ is denoted by z_ρ. The potential φ_ρ of such a dipole configuration is thus analytically known. Because it depends on the actual position bftor $\mathbf{r}_\rho$ it has to be sampled during the run, too.

For a rough estimate of the correction term take the fixed ion positions as reference points for the value of the potential and expand the remainders of the sum of all four terms in equation 7, viz.

$$\sum_{i,\rho} [-\varphi_\rho(Z_i - z_\rho) + \varphi_\rho(z_i - z_\rho) - \varphi_\rho(z_i - Z_\rho) + \varphi_\rho(Z_i - Z_\rho)] =$$

$$\sum_{i,\rho} [-\varphi_\rho''(Z_i - Z_\rho)(z_i - Z_i)(z_\rho - Z_\rho)] = \qquad (11)$$

$$- \sum_i \varphi_\rho''(0)(z_i - Z_i)^2 = -n_0 N <z^2> \qquad (12)$$

with $\varphi_\rho''(0)$ equal to the negative density at the central layer which involves an additional minus sign because of the above definition of the correction term, thus equal to n_0. An averaging over the positions of the N electrons leaves only diagonal terms yielding the average of $<z^2>$ which depends quadratically on the layer spacing and is independent of the jellium extent z_0, supposed $z_0/2$ lies significantly outside the ion layers. This is the correction to be added to the energy afterwards if an exact sampling of it with the hamiltonian shall be avoided. A correction also arises in the exponent of the Jastrow factor where the two-body repulsion is summed and tabulated in the same manner as in equation 10. An approximate short cut treatment is not possible there, however, because no cancellation can occur as it is present in the hamiltonian between ions and electrons. Instead, the correction may be taken into account through the minimum property of the wave function and the variational parameters, especially through an optimum choice of z_0 in the Jastrow factor.

In figure 4 a specific example is shown for minimizing the total energy of a slab with the ideal GaAs(110) surface. As discussed above, the surface system needs an

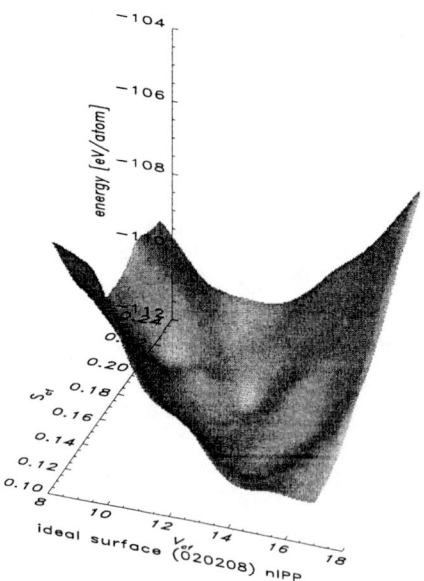

FIGURE 4. Total energy vs. slope, S_{cf}, and height, V_{cf}, of additional one-body function in Jastrow exponent to compensate for overshooting repulsion by two-body term in case of GaAs(110) slab with 2*2*8 layers.

additional one-body function to reduce the electron repulsion caused by the two-body Jastrow term. An energy surface is shown depending on the two essential parameters of that function of z, i.e. the slope and the maximum height. One observes a discernible variation on that energy scale with a clear minimum. The overall global minimization is finally done by known statistical optimization, where a fit surface is adapted to the set of QMC data within a controlled accuracy, and the optimum values are derived. As a result, surface relaxations and reconstructions can be resolved, see figure 5 where an energy minimization of the dangling bond direction is displayed. Thus, an optimized wave function in analytic form is available through this method which can be used as a ground state for the purpose of electron spectroscopy.

C QMC in photoemission

There are several stages where QMC could be of value within and beyond one-step photoemission calculations, e.g.

- determination of the ground state as a many-body wave function,
- many-body transport calculations for the outgoing emitted state of the detected electron,

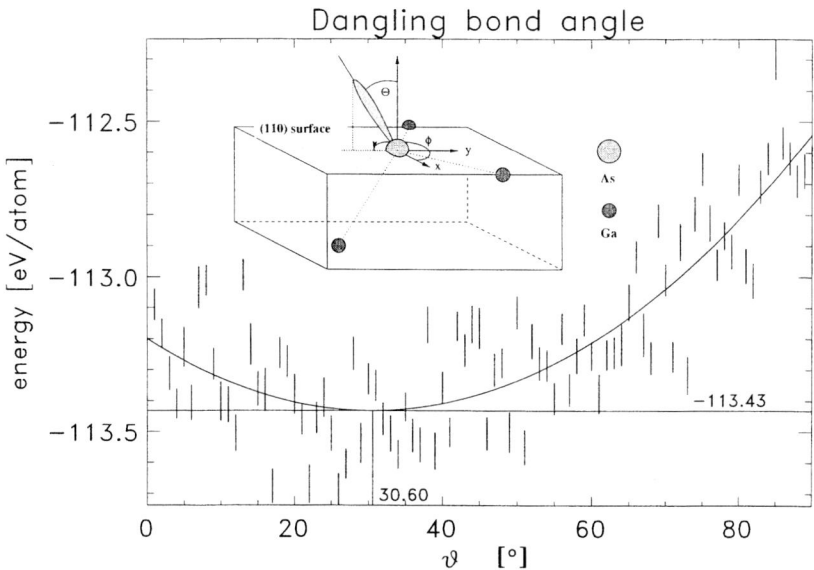

FIGURE 5. Total energy vs. angle of direction of dangling bond hybrid (inset), as formed in one-body wave function of Slater determinant in case of GaAs(110) slab.

- evaluation of the light excitation process with the foregoing states, and the vertex corrections.

The first item has been seen to be feasible in that total energy minimizations can be carried through for semiconductor surface determinations, and that they yield statistically significant results. Actually, the application to a multitude of structures is not advised because of the necessity of large computing resources. However, extracting many-body effects by sampling the electron correlation function is in progress and will be used to construct a surface specific density functional. Subsequent DFT calculations lead to one-particle states to be introduced into the photoemission golden-rule formula. This certainly improves the actual procedure in photoemission calculations, but does not represent the ultimate goal which could be expected from a many-body treatment. In contrast, this might be hoped to arise from a joined many-body evaluation of the whole photoemission process by QMC methods.

The transport topic has a few predecessors in high-field transport of semiconductors where QMC techniques have been tentatively applied, essentially with the path integral. [29] Several methods to attack this problem have been investigated. [15] The central quantity would be the scattering matrix between an incoming state, whatever state may be formulated being left by the light excitation, and a free electron state with momentum **k** as detected, together with the N-body wave function of the remaining electrons. In an approximation similar to the "blue electron" idea [30], the photoelectron is separated from the rest if its one-particle wave function is treated outside the determinant, e.g. a plane wave multiplied with the N-determinant. In generalizing, exchange (correlation) of the photoelectron with the system could be included by introducing the free wave into the determinant (Jastrow factor), respectively. The incoming state similarly is decomposed into a N- and a one-particle state, the latter further decomposed into a set of plane waves of momentum $\mathbf{k}'$. Thus, the scattering matrix $\chi_{kk'}(t)$ between two plane waves, both multiplied by the same N-particle determinant representing the electron background has to be calculated.

$$\chi_{kk'}(t) = \int dr_0^3...dr_N^3 \Psi^*_{N+1,k'} e^{-iHt} \Psi_{N+1,k} \quad (13)$$

$$\Psi_{N+1,k} = \Psi_N e^{i\mathbf{k}\mathbf{r}_0} \quad (14)$$

Because of the many-particle hamiltonian in the time evolution this is a real many-body problem even with the assumed decoupled states. It can be cast into small time steps according to the Trotter formula, which yields a product of a large number of single time step matrices to be integrated over all intermediate states such as in the usual path integral, viz.

$$e^{-iH\tau} \longrightarrow e^{-iH_N\tau} e^{-iW\tau} e^{-iH_1\tau}, \quad \text{for} \quad \tau \to 0, \quad (15)$$

$$\chi_{kk'}(t) = \sum_{k_1...k_{P-1}} \chi_{kk_1}(\tau) \chi_{k_{P-1}k'}(\tau) \prod_{j=1}^{P-2} \chi_{k_j k_{j+1}}(\tau) \quad (16)$$

with hamiltonian

$$H = H_N + W + H_1, \quad (17)$$

$$H_1 = -\frac{1}{2}\Delta_0 + V(\mathbf{r}_0). \quad (18)$$

consisting of the N-particle part H_N, the hamiltonian of the photoelectron H_1, and the interaction W between both. The single step scattering matrix is

$$\chi_{kk'}(\tau) \sim e^{-\frac{\tau}{4}(k^2+k'^2)} \int dr_0 <N|e^{-iW(\mathbf{r}_0)\tau}|N> e^{-iV(\mathbf{r}_0)\tau} e^{i(\mathbf{k}-\mathbf{k}')\mathbf{r}_0}. \quad (19)$$

The first step in equation 19 consists in sampling the time evolution by the interaction W for various positions $\mathbf{r}_0$ of the photoelectron. The result is given by a subsequent Fourier transform to be inserted into equation 16, whose evaluation is a crucial step from computational demands. Therefore, the time step length cannot be chosen too small on one hand, but must be small enough to guarantee the valdity of the Trotter formula. Tests on the usefulness of equation 16 were performed for one-particle states. The MC evaluation of the time propagator yielded strongly fluctuating results, and even special procedures to directly handle alternating series did not overcome the problem of an uncontrollable increase of rounding errors.

Instead, it was found [31,32] that a reasonable procedure consists in expanding the time propagator into Chebyshev polynomials, viz.

$$e^{i\hat{H}t} = \sum_{j=0}(2 - \delta_{j,0})J_j(R)\phi_j(i(\hat{H}t - S)/R) \qquad (20)$$

with suitable c-numbers R, S, and the iteration procedure

$$\Phi_j = 2\hat{H}\Phi_{j-1} + \Phi_{j-2}. \qquad (21)$$

An exponential convergence is expected for this series. The formalism was applied first to a one-particle system incorporating an additional static electric field in the potential V of equation 18. The purpose was to investigate also high field transport as a separate topic with this method. Starting with a plane wave state $|\psi> = |\mathbf{k}>$, the repeated application of the hamiltonian reads for a single step

$$\hat{H}|\psi> = \sum(|\mathbf{r}> V(\mathbf{r}) <\mathbf{r}|\psi> + |\mathbf{k}> k^2/2 <\mathbf{k}|\psi>, \qquad (22)$$

where the crystal and external potential are treated in direct space, and the kinetic energy in Fourier space. An empirical pseudopotential [33] of GaAs was used for the ease of testing. The time development was started with a Bloch state of this potential. An optical potential had to be introduced which represents the inelastic losses which in the future development will be taken into account through the many-body interactions supplied by a coupling to thermostats. Here, it was chosen as $\tilde{V}_{opt} = -\frac{1}{2}k_0\sqrt{\bar{p}^2}$ with $k_0 = 1/$(mean free path) to reflect the saturation of the energy gain of the electron in the electric field by introducing a dependence on the velocity fluctuation operator. The rapidly converging procedure showed results such as plotted in figures 6, 7. The calculations did not show any instabilities.

On a short time scale of femtoseconds, see figure 6 one observes a saturation of the expectation values of velocity and energy. The characteristic time depends on the value of the damping constant k_0, and increases with decreasing damping as to be expected. [15] This first variation, after switching on, may be considered as a fluctuation forced by the initial values. After that, the long time scale drift sets in, which is depicted in figure 7. Thereby, the electron reaches velocities which correspond within the order of magnitude to the gain of energy according to a free particle dispersion law. Further time evolution shows the oscillations between Brillouin zone boundaries, in energy as well as in velocity. The oscillations slow down in the period by a factor of two and are smoothed if the optical potential is switched off. [15] The effect of the special choice of the optical potential, such as the above dependence on the fluctuations, is not fully understood and deserves further consideration. The first steps in a many-body simulation could take advantage of using a suitable optical potential, i.e. an approximation to the self-energy of a quasi-particle. Summarizing, the time development as it occurs in the scattering states of photoemission can be adequately simulated by a Chebyshev expansion which allows a rapidly convergent transition to long time scales. It might even make the use of the Trotter formula together with the path integral formulation obsolete for this purpose in favour of direct time integration.

The final step in a statistical many-body treatment is represented by the simulation of the light excitation process. It should incorporate the elastic and inelastic

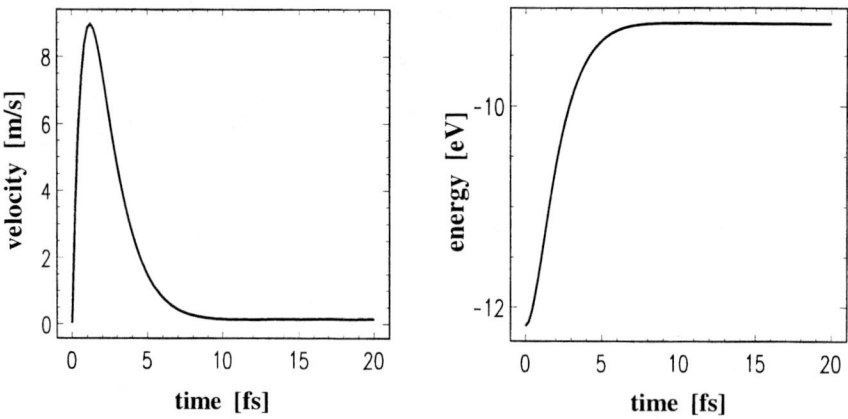

FIGURE 6. Development on short time scale of one-particle Bloch state at Γ in the lowest valence band of GaAs with damping $k_0 = 75$ a.u. driven by electric field $E = 100$ kV/cm, average velocity (left) and average energy (right).

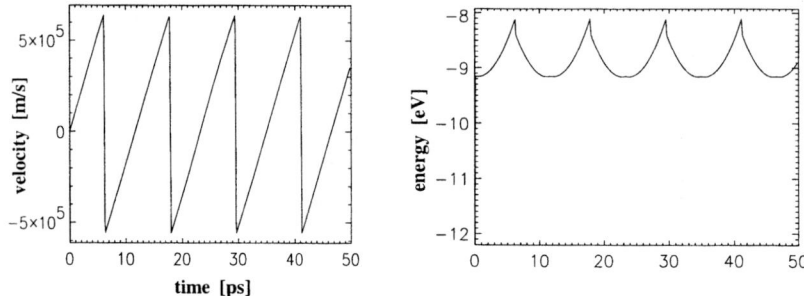

FIGURE 7. Development on long time scale, see caption of figure 6.

processes, intrinsic as well as extrinsic ones, and should especially take care of the presence of the hole and its interaction with the photoelectron. This is the least understood part of the whole development and it would be mostly speculated to go into details here. However, holes can be introduced into the determinant, or perhaps more general into several determinants, as missing states, and are to be left out in the Jastrow factor. The transition from the well investigated ground state to such a scattering state would be an suitable concept if the numerical implications can be solved. As indicated above, a path exists which yields the scattering states and on which one could gradually advance to higher reliability with respect to the many-body effects.

CONCLUSIONS

Presently, the one-step model yields the most accurate technique, to calculate photoemission spectra. Especially, in the case of valence bands this tool is indispensable for a reliable interpretation of experimental spectra. One can take into account several many-body effects with this method, mainly ground state spectral density, scattering state's self-energy, and photon renormalization by the solid's response. However, all these can be considered as self-energy insertions, whereas the problem of vertex corrections, e.g. electron-hole interactions, is principally different and far from being solved.

In both respects, self-energy and vertex corrections, QMC will not only supply additional information, but partly yield fundamental knowledge, as it is well accepted for the first topic in the case of DFT. QMC simulations for the photoemission process are a promising field, and a basic procedure seems to be clear, even if the numerical details have not yet been fully established. However, the ground state part can be considered as being solved for simple systems, the scattering states can be principally obtained by the Chebyshev procedure as sketched here. Together with a generalization of the one-determinant ansatz to a consideration of several configurations these demand to be the next, most important tasks towards a many-body photoemission theory.

The bottleneck, from the author's knowledge, is represented by the available computing resources, obvious in those cases where the codes have already been fully developed and could be extensively applied. All of the surface calculations have been performed on a T3E parallel computer with an almost linear scaling.

ACKNOWLEDGMENT

Parallel-computer facilities provided by the Konrad-Zuse-Zentrum für Informationstechnik, Berlin, and the John von Neumann-Institut für Computing, Jülich are gratefully acknowledged. The author is indebted to R. Bahnsen and T. Strasser for discussions and help in preparing the manuscript. The work is supported by the Deutsche Forschungsgemeinschaft within the project No. Scha 360/17-1.

REFERENCES

1. D. M. Ceperley, B. J. Alder, Phys. Rev. Lett. **45** (7), 566 (1980)
2. G. Sugiyama, G. Zerah, B. J. Alder, Physica A **156**, 144 (1989)
3. H. Eckstein, W. Schattke, Physica A **216**, 151 (1995)
4. G. Yao, J. G. Xu, X. W. Wang, Phys. Rev. B **54**, 8393 (1996)
5. S. Fahy, X. W. Wang, S. G. Louie, Phys. Rev. Lett. **61** (14), 1631 (1988); S. Fahy, X. W. Wang, S. G. Louie, Phys. Rev. B **42** (6), 3503 (1990)
6. A. J. Williamson, S. D. Kenny, G. Rajagopal, A. J. James, R. J. Needs, L. M. Fraser, W. M. C. Foulkes, P. Maccullum, Phys. Rev. B **53**, 9640 (1996)
7. H. Eckstein, W. Schattke, M. Reigrotzki, R. Redmer, Phys. Rev. B **54**, 5512 (1996)
8. A. Malatesta, S. Fahy, G. B. Bachelet, Phys. Rev. B **56**, 12201 (1997)
9. R.Q. Hood, M. Y. Chou, A. J. Williamson, G. Rajagopal, R. J. Needs, W. M. C. Foulkes, Phys. Rev. Lett. **78**, 3350 (1997)
10. Williamson, R. Q. Hood, R. J. Needs, G. Rajagopal , Phys. Rev. B **57**, 12140 (1998)
11. W.M.C. Foulkes, R.Q. Hood, R.J. Needs, Phys. Rev. B **60** 4558 (1999)
12. P. R. C. Kent, R. Q. Hood, M. D. Towler, R. J. Needs, G. Rajagopal, Phys. Rev. B **57**, 15293 (1998)
13. P.R.C. Kent, R.Q. Hood, A.J. Williamson, R.J. Needs, W.M.C. Foulkes, G. Rajagopal, Phys. Rev. B **59**, 1917 (1999)
14. R. Bahnsen, D. Schulz, W. Schattke, R. Redmer, Czech. J. Phys. **49**, 1519 (1999); R. Bahnsen, W. Schattke, R. Redmer in preparation
15. H. Eckstein, PhD thesis, Kiel (1996); available at http://www.tp.cau.de/schattke
16. J.Henk, W.Schattke, Comput. Phys. Commun. **77**, 1 (1993)
17. A.Bödicker, W.Schattke, J.Henk, R.Feder, J. Phys.: Condens. Matter **6**, 1927 (1994)
18. W. Schattke, Prog. Surf. Sci. (2000) in press; available at http://www.tp.cau.de/schattke
19. W. Schattke, Prog. Surf. Sci. **54**, 211 (1997); W. Schattke, Web-Proceedings of the ALS Workshop on *Theory and Computation for Synchrotron Applications*, ed. M.A. Van Hove (Berkeley CA, 1997), available at URL: http://electron.lbl.gov-/alsworkshop/proceedings/index.html and at http://www.tp.cau.de/schattke
20. T. Strasser, C. Solterbeck, F. Starrost, W. Schattke, Phys. Rev. B **60**, 11577 (1999)
21. S.S. Dhesi, C B. Stagarescu, K E. Smith, D.Doppalapudi F.Singh, T D. Moustakas, Phys. Rev. B **56**, 10271 (1997)
22. C. Solterbeck, W. Schattke, J.-W. Zahlmann-Nowitzki, K.-U. Gawlik, L. Kipp, M. Skibowski, C.S. Fadley, M.A. Van Hove, Phys. Rev. Lett. **79**, 4681 (1997)
23. R. Herrmannn, S. Samarin, H. Schwabe, J. Kirschner, Phys. Rev. Lett. **81**, 2148 (1998)
24. S. Ogawa, H. Nagano, H. Petek, A.P. Heberle, Phys. Rev. Lett. **78**, 1339 (1997)
25. G. Ortiz, Phys. Rev. B **45**, 11328 (1992)
26. M. Causà, A. Zupan, Chem. Phys. Lett. **220**, 145 (1994)
27. B. Paulus, P. Fulde, H. Stoll, Phys. Rev. B **54**, 2556 (1996)
28. N. Metropolis, A.W. Rosenbluth, M.N. Rosenbluth, A.H. Teller, E. Teller, J. Chem. Phys. **21** (6), 1087 (1953)
29. B.A. Mason, K. Hess, Phys. Rev. B **39**, 5051 (1989)

30. L. Hedin, Nucl. Instrum. Methods Phys. Res. A **308**, 169 (1991)
31. H. Tal-Ezer, R. Kosloff, J. Chem. Phys. **81**, 3967 (1984)
32. R. Kosloff, Ann. Rev. Phys. Chem. **45**, 145 (1994)
33. M.L. Cohen, T.K. Bergstresser, Phys. Rev. **141**, 789 (1966)

Photoemission Revealing Signature of Stripes and Orbital Modulation in the High T_c Superconductors

N. L. Saini and A. Bianconi

Unitá INFM and Dipartimento di Fisica, Università di Roma "La Sapienza"
P.le Aldo Moro 2, 00185 Roma, Italy

Abstarct. Here we report use of angle resolved photoemission (ARPES) in the angle scanning mode to explore the Fermi surface topology of the high T_c superconductors. We have measured momentum distribution of the spectral weight at the Fermi level of $Bi_2Sr_2CaCu_2O_{8\pm\delta}$ (Bi2212) superconductor. The measurements were made in two different experimental geometry to take care of the polarization dependence of the transition probability (matrix element effects). We find an asymmetric suppression of spectral weight around $(\pi,0)$ that is a direct indication of coupling of the itinerant carriers with an incommensurate charge density wave (ICDW) or with an 'orbital wave', i.e. stripes, with a wave vector $q(0.4\pi, 0.4\pi)$.

INTRODUCTION

Even after a decade we are away from understanding of the Fermi surface of high T_c superconductors retaining several controversies including the nature of the electronic states and their behavior with doping and temperature. The representative system to reveal the Fermiology of the cuprates has been the $Bi_2Sr_2CaCu_2O_{8\pm\delta}$ (Bi2212) material, because of its technical suitability for the angle resolved photoemission (ARPES) that is the recognized technique for revealing fundamental electronic properties of the high T_c superconductors.

One of the fundamental parameters to understand the physics of superconductors is the coupling between the electrons at the Fermi surface, i.e., the electron-electron interaction $V_{kk'}$. It is thus vital to obtain the information on the distribution of wave-vectors k of the electrons at the Fermi surface. In the past, there has been a number of photoemission studies revealing several features such as saddle point singularity, and pseudogaps, destruction of Fermi surface below T_c. In the superconducting phase a d-wave symmetry for the superconducting gap has been found (1, 2). However, in most of these studies the standard energy distribution curves (EDC) were measured. Topological information i.e., the momentum distribution of electrons at the Fermi surface of the Bi2212 superconductor was first measured by Aebi et al (3) with

ARPES in the angle scanning mode (4) using unpolarized laboratory source. Indeed this study indicated some new features at the Fermi surface such as shadow bands. However, a breakthrough came in the field when the complete topology of the Fermi surface was measured using polarized synchrotron radiation light with photon energy of 32 eV (5). The study made clear the actual segments and missing parts of the Fermi surface including the spectral weight suppression by the pseudogap triggering several new studies to revisit the Fermi surface topology of the Bi2212 system (6-8). The suppression of spectral weight near the M points was shown to be due to an incommensurate charge density wave with a wavevector $q(0.4\pi, 0.4\pi,)$ (5, 9).

The new topological features of the Fermi surface revealed by the experiments of Saini et al (5, 9), with photon energy of 32 eV, were later confirmed by Y. -D. Chuang et al (6), who measured momentum distribution of the electrons at the Fermi surface using EDC, by Feng et al. (7), who gave a systematic look to the momentum distribution topology, and by Fretwell et al (8) who extended the momentum distribution experiments below T_c, reproducing the topology of the M point in the superconducting state with higher momentum and energy resolution. It should be recalled that the topology of the electron momentum distribution around the M point is one of the most relevant features due to the fact that the superconducting condensate and the opening of the pseudogap appear to reside in regions near the M point while in the other high symmetry directions the Fermi surface remains ungapped (10, 11).

In this contribution we report an extension of our work on the topological features of the Fermi surface of the Bi2212 system measured using polarized synchrotron radiation light with photon energy of 32 eV. We have investigated the momentum distribution of electronic states near the Fermi level within an energy of 50 meV, considering the fact that the pairing of the electrons near the Fermi level, interacting with collective excitations such as phonons or charge/spin density waves, within a characteristic cut off of the order of 50 meV. We have identified the asymmetry of the spectral weight suppression around the $M(\pi,0)$ point, due to the coupling of itinerant carriers with an incommensurate charge density wave (ICDW) with wavevector in the diagonal direction $q(0.4\pi, 0.4\pi)$. The polarization dependence of the matrix element of photoemission transition probability has allowed us to further discriminate various features. Finally we argue that the energy dependent matrix element effect could be assigned to 'orbital waves' associated with the ICDW.

EXPERIMENTAL DETAILS

Angle resolved photoemission measurements were made at the Laboratoire pour l'Utilisation du Rayonnement Electromagnetique (LURE) (Orsay-France) on the SU6 beamline. A single crystal ($T_c \sim 91K$) of size $6 \times 6 \times 1$ mm^3, grown by floating zone method, was used for the experiments made in the EDC mode and the angle scanning mode. Sample alignment for these measurements is vital part as a small misalignment could confuse the identification of the real features. The large crystal used for the measurements was aligned by specular laser reflection from the crystal surface. The clean and flat surface was obtained by cleaving the crystal in-situ at room temperature.

The experiments were performed in an Ultra-High Vacuum (UHV) chamber (base pressure 7 x 10^{-11} mbar) equipped with an angle-resolved hemispherical analyzer and a high precision manipulator which permits rotation in the full 360° azimuth emission angle (ϕ) and 90° polar emission angle (θ) relative to the surface normal (12). The photoelectron intensities at the Fermi level and bellow E_F were recorded along a series of azimuth scans. The sample was rotated around its normal and the intensity was recorded every 1.5° at fixed theta, with an absolute angular precision of better than 0.5°. The experiments were repeated with different cleavage, each cleave gave the same results showing the high quality of the crystal with stable surface for several days.

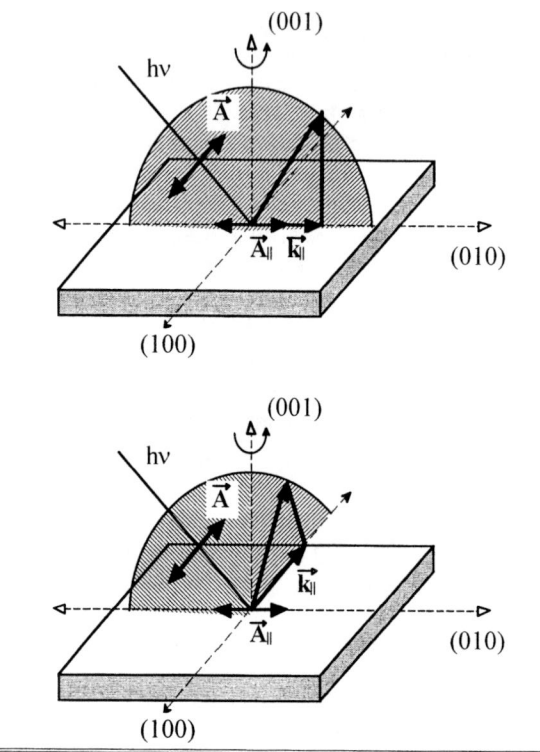

FIGURE 1. Pictorial view of the experimental geometry used for the measurements. The upper (lower) panel shows the "even" ("odd") geometry. The crystal axes of an orthorhombic lattice (Bi2212 system) are indicated respectively by (100), (010) and (001). The detector movement to change photoelectron wave vector is within (out of) the scattering plane for the "even" ("odd") geometry.

The constant energy (Fermi surface) contours were obtained by centering the electron energy window at the E_F and collecting the electrons within an energy window of the order of spectrometer resolution (~50 meV) using a photon energy of 32 eV. The polarization vector of the synchrotron light, the direction of the photon beam and the surface normal were kept in the same horizontal plane, called "scattering plane", for all the measurements. The "mirror plane" is defined by the sample normal and the direction of the emitted photoelectron selected by the detector position. The detector was moved in the fixed "mirror plane", by changing the polar angle, in order to select different values of the wave vector in the superconducting CuO_2 plane ($k_\parallel$). The direction of the initial state $k_\parallel$ was selected by rotating the sample around its normal that was ascertained to be collinear with the crystallographic c-axis.

The Fermi surface contours were measured in two different polarization geometry shown pictorially in Fig. 1. The upper picture shows the "even" geometry while the lower picture represents the "odd" geometry. For the "even" geometry, the 'scattering plane' is coplanar to the 'mirror plane' while for the "odd" geometry the 'mirror plane' is orthogonal to the 'scattering plane' (1,2,13).

The photoemission signal is proportional to the photo-ionization cross-section given by the Fermi Golden rule

$$\frac{d\sigma}{d\Omega} \propto \left| \langle \Psi_f(k) | A \cdot p | \Psi_i(k) \rangle \right|^2 \delta(E_f - E_i - h\nu)$$

where **p** is the momentum operator and **A** the vector potential of the photon beam collinear with the photon electric field of the energy $h\nu$, $\Psi_f(k)$ is the final state at energy E_f, $\Psi_i(k)$ is the initial state at energy E_i. The matrix element must be invariant under the geametrical operations for the measurements since the only combinations of initial state, final state and vector potential those leave the matrix element invariant will contribute to the photoemission signal.

The momentum operator **A·p** has the same symmetry as **A·r** (**r** being the direction of the emitted photoelectron). For the "even" ("odd") geometry the momentum operator has even (odd) symmetry with respect to the mirror plane, in fact, the electric field lies within (out-of) the mirror plane. Therefore for the "even" ("odd") geometry only initial state wave functions of even (odd) symmetry with respect to the mirror plane are accessible. The initial states of b_1 symmetry (formed by a mixing of Cu $3d_{x^2-y^2}$ and O $2p_{x,y}$ orbitals) are even with respect to the mirror plane for k_i parallel to the Cu-O-Cu bonds while they are odd with respect to the mirror plane for k_i at $45°$ to the Cu-O-Cu bonds. Thus in the "even" experimental geometry the transition from these states forming the conduction band is fully allowed for k_i in the Γ-M $(0, \pi)$ and equivalent directions (the Cu-O-Cu direction) and forbidden for k_i in the Γ-X (π, π) and equivalent directions (Cu-Cu direction). On the contrary, in the "odd" geometry the transition from the states is fully allowed for k_i in the Γ-X (π, π) and equivalent directions and forbidden for k_i in the Γ-M $(0, \pi)$ and equivalent directions.

RESULTS AND DISCUSSION

The constant energy contours measured at the Fermi energy in the two polarization geometry are shown in Fig. 2. The upper panel shows the Fermi surface image obtained using the so-called "odd" experimental geometry while the lower panel represents the image obtained in the "even" experimental geometry (see e.g. Fig. 1).

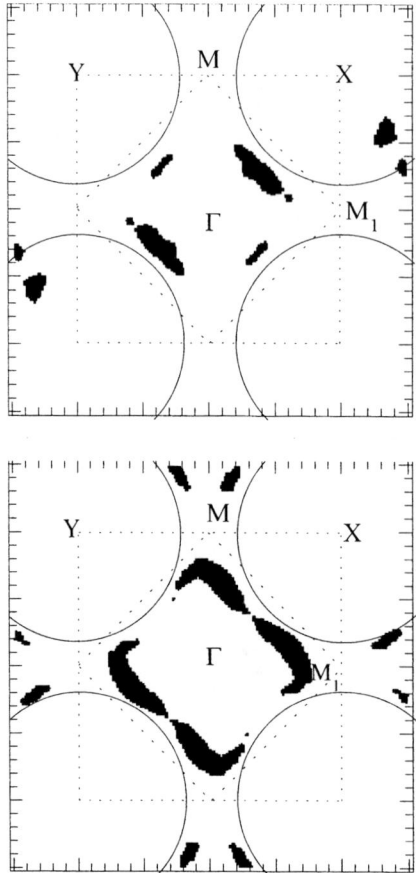

FIGURE 2. Fermi surface contours measured on the Bi2212 system with "even" (lower) and "odd" (upper) geometry. The photointensity is given in grey scale, increasing from light to dark. The wavevector in the CuO_2 plane is given in units of π/d, (d is Cu-O-Cu distance = 3.8Å) and the notations define the high symmetry points in the Brillouin zone; $\Gamma(0, 0)$, $M_1(\pi, 0)$, $M(0, \pi)$, $X(\pi, \pi)$ $Y(-\pi, \pi)$, where Γ–M is along the Cu-O-Cu bond direction.

The photoemission intensity is given by grey colours with the dark regions indicating higher intensity of emitted photoelectrons excited from the initial state having constant energy E_F and in-plane wavevector ($k_\parallel$) spanned over the reciprocal space of the two dimensional CuO_2 plane. There is clear photointensity suppression along the Γ–M and equivalent directions in the upper image due to matrix element effects as discussed above. Similarly, the photoemission intensity is suppressed along the Γ-Y and Γ-X direction for the image measured in the "even" experimental geometry (lower panel) as the transition from the initial states having odd symmetry with respect to the mirror plane is not allowed. Therefore, in principle, the combination of the two polarization provides an opportunity to clearly identify the real topology of the Fermi surface in all high symmetry directions, i.e., the image obtained in the "odd" experimental geometry reveals clear information about the emission along the two orthorhombic directions (Γ-X and Γ-Y) while the "even" experimental geometry allows to identify the real momentum distribution of the electrons at the Fermi surface in the tetragonal directions (Γ-M and equivalent directions).

One of the anomalies in the Fermi surface topology could be seen by looking at the photointensity along the two high symmetry directions, Γ-X and Γ-Y. The selection rules (13) allow maximum emission along the two directions in the image measured using the "odd" geometry and minimum emission in the image measured with the "even" geometry. In addition, the photointensity along the two directions is expected to be equivalent for a lattice symmetry with initial states having b_1 symmetry (Cu $3d_{x^2-y^2}$ - O $2p_{x,y}$). However, this is not the case, and asymmetry of the photointensity could be clearly seen in the global images shown in Fig. 2. In fact a clear photointensity difference is visible in the upper image along the Γ-X and Γ-Y directions while the matrix element should be the same with initial states having only Cu $3d_{x^2-y^2}$ orbitals ($m_l=2$) lobbing towards the O 2p orbitals. This broken topological symmetry could be better seen in the lower panel showing the image measured using the "even" geometry. Indeed, the photointensity is absent along the Γ-Y direction, as expected from the selection rules, however, a clear photointensity is observed along the Γ-X direction.

To further investigate we have measured independently the photointensity along the Y(-π, π) and X (π, π) directions in the two experimental geometry and shown in Fig. 3. Again the upper panel represents the "odd" geometry while the lower panel shows the photointensity obtained with the "even" geometry. In the upper panel a broad peak around 0.35 2π/a could be seen due to the main band while a peak around 0.65 2π/a represents a contribution of shadow features which appear around (0.5π, 0.5π) and (-0.5π, 0.5π) due to small hole-pockets expected in lightly doped Mott-insulators (3, 14). It is evident that the photointensity due to the main band crossing is different in the two directions, being higher in the X (π, π) direction. Similarly, the photointensity scans in the two directions measured in the "even" geometry (lower panel) shows an evident asymmetry of the Fermi surface topology.

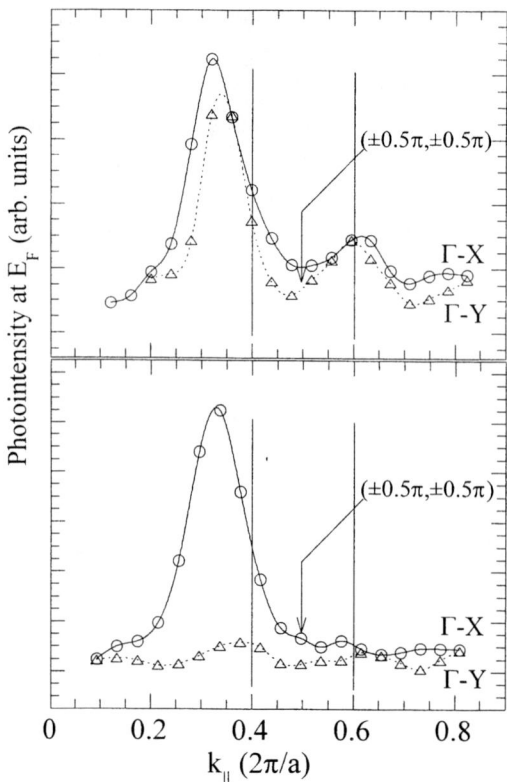

FIGURE 3. Photointensity along X (solid line) and Y directions (dotted line) with "even" (lower) and "odd" (upper) geometry.

The shadow bands and shadow features at the Fermi surface were first seen by Aebi et al in photoemission study made in the angle scanning mode (3). The observation of shadow features was followed by several other groups and some signatures of these were seen also in the energy distribution curves (15), however, presence of these features have been always an issue of controversy. The present measurements provide a definitive proof for these features at the Fermi surface which could be possible because of fully allowed transition in the first Brillouin zone along the (π, π) and $(-\pi, \pi)$ directions in the "odd" geometry. In addition, the contribution from the shadow bands could be well differentiated along the (π, π) direction that is free from any overlapping umklapp bands (weak features appears due to scattering of the exiting photoelectrons with the modulation of the Bi-O plane (2)). Moreover, the direct photointensity scans along the two orthorhombic directions show that the contribution of shadow features is significant. The shadow features on the Fermi surface are expected in systems with antiferromagnetic correlation giving a shifting of the spectral function by the antiferromagnetic wavevector $G(\pi, \pi)$. This gives hole pockets closed around $(0.5\pi, 0.5\pi)$. Thus the shadow bands observed in this work might be due to antiferromagnetic correlations lasting upto higher doping (3,14). It has also been argued that these features may appear due to orthorhombic symmetry of the Bi2212 system giving folding of the Brillouin zone (16).

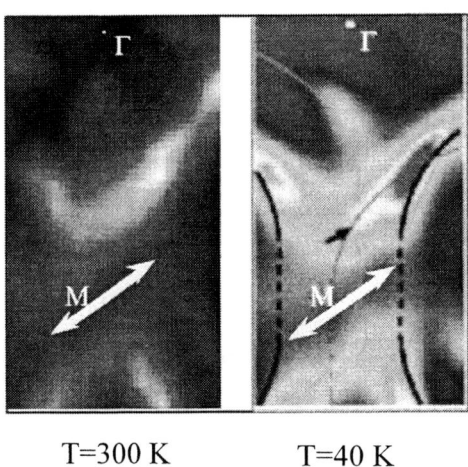

T=300 K   T=40 K

FIGURE 4. Topology of the $(\pi,0)$ point at the Fermi surface (ref. 5) measured in the normal state (left) is compared with the one measured by Fretwell et al in the superconducting state (ref. 8). The diagonal wavevector $q(0.4\pi,0.4\pi)$ is shown. The yellow colour represent maximum photointensity for a minimum corresponding to the black colour in the left panel while the photointensity maximum corresponds to the red colour with a minimum corresponding to the sky blue colour in the right panel of the figure. The two images clearly demonstrates evolution of the spectral weight at the Fermi surface around the M point across the superconducting transition.

Let us turn to the topology of the M point at the Fermi surface of the Bi2212 system. The initial states of b_1 symmetry (formed by a mixing of Cu $3d_{x^2-y^2}$ and O $2p_{x,y}$ orbitals) are even with respect to the mirror plane for the k_i parallel to the Cu-O-Cu bonds, i.e. along the Γ-M direction in the "even" experimental geometry and hence the image shown in the lower panel of Fig. 2 represents real topology of the M $(\pi,0)$ points. We can see a strong suppression of spectral photointensity around the $(\pi, 0)$ and equivalent locations at the Fermi surface. This asymmetric suppression is confirmed by Chuang et al (6), Feng et al (7) and Fretwell et al (8) in recent experiments. After a careful analysis we have identified a well defined wavevector $q(0.4\pi, 0.4\pi)$ over which the suppression occurs (9). We have shown the topology of the Fermi surface around the $M_1(\pi,0)$ point in Fig. 4. The topology obtained in the normal state (5) is compared with the topology of the M_1 point in the superconducting state measured by Fretwell et al (8) in recent experiments. The agreement between the two data sets is indeed impressive.

It should be recalled that the photointensity suppression around the M point is less prominent at lower photon energy of 22 eV that has claimed to be due to energy dependence of the photoemission matrix element (17). In fact, it is clear from the recent experiments that the Fermi surface of the Bi2212 is more complex than it was thought and indeed the matrix element effects play vital role in its measurements. However, a simple matrix element effect gives a symmetric suppression of the photoemission intensity around the M point (17) rather than asymmetric as observed in the present experiments. Earlier we have shown that the asymmetric suppression around the M points is a clear indication of the coupling of the itinerant carriers with an incommensurate charge density wave (ICDW) with a wave vector $q(0.4\pi, 0.4 \pi)$ i.e., stripes. Indeed, there are several experiments providing evidence for the stripe instability where a polaronic charge density wave coexist with itinerant charge carriers in the cuprates (18-25). In the Bi2212 system, the stripes within the CuO_2 have been shown by Cu K-edge extended x-ray absorption fine structure (EXAFS) and anomalous diffraction measurements (25) In fact, the photointensity suppression is over a well defined nesting vector $q(0.4\pi, 0.4\pi)$, that is the second harmonic of the main lattice modulation.

In the cuprates the polarons are indeed of the Q_2-type (so-called a d_{z^2} polaron) (26). The characteristics of the polaronic charge density waves involve modulation of the orbital angular momentum, i.e., a modulation of the mixing of the $m_l=0$ (and $m_l=1$) component with the majority $m_l=2$ component due to local displacements involving rhombic distortions of the CuO_4 square plane, breaking the local crystal symmetry. The coupling of itinerant electrons with this 'orbital wave' is expected to influence the spectral weight at the Fermi surface. In fact, the suppression due to an 'orbital wave' is expected to be dependent on the photon energy via matrix element effects. Thus, the fact that the anisotropic suppression, connected by the $q(0.4\pi, 0.4 \pi)$, appears photon-energy-dependent and the asymmetric suppression exists also in the superconducting state, it constructs an experimental ground for coexistence of 'orbital waves' and superconductivity. In fact the orbital modulation is associated with modulated distortions of the CuO_2 unit cell. This will also explain the asymetry of the first Brillouin zone with different spectral weight along the Γ-X and Γ-Y directions (Fig. 2 and Fig. 3), the asymetry that could not be accounted by matrix element effects considering a perfect tetragonal symmetry of the unit cell with only $d_{x^2-y^2}$ orbitals ($m_l=2$).

In summary, a complete topology of the spectral weight has been measured near the Fermi level of Bi2212 superconductor. The measurements have been made in two polarization geometry allowing us to differentiate the real topological features at the Fermi surface. The results show a clear asymmetry of the spectral weight at the Fermi surface. Asymmetric spectral weight suppression around the M points is observed. We have discussed that the asymmetric suppression is related with an incommensurate charge density wave with a wavevector $q(0.4\pi, 0.4\pi)$, i.e., stripes, involving modulation of the orbital momentum. The results have also shown a clear proof for the presence of the shadow bands at the Fermi surface. In conclusion, the suppression of the spectral weight at the M points, associated with the pseudogap, is related to the stripe fluctuations in the system, i.e., pseudogap is associated with stripe minigaps (27, 28), and the electrons moving in the stripes involve interaction with an 'orbital wave'.

ACKNOLEDGEMENTS

This work was partially funded by Istituto Nazionale di Fisica della Materia (INFM) and Consiglio Nazionale delle Ricerche (CNR). The authors would like to thank A. Lanzara for invaluable collaboration in this work. The authors also thank J. Avila and M. C. Asensio for the collaboration in photoemission measurements and S. Tajima, G. D. Gu and N. Koshizuka in the crystal growth and characterization.

REFERENCES

1. see e.g. a review, Shen, Z.-X. and Dessau, D. S., *Physics Reports* **253**, 1 (1995) and references cited therein.
2. see e.g. for a review, Randeria, M and Campuzano, J.C., Varenna Lectures 1997, cond-mat/9709107.
3. Aebi, P., Osterwalder, J., Schwaller, P., Schlapbach, L., Shimoda, M., Mochiku, T. and Kadowaki, K., *Phys. Rev. Lett.* **72**, 2757 (1994); Osterwalder, J., Aebi, P., Schwaller, P., Schlapbach, L., Shimoda, M., Mochiku T., and Kadowaki, K., *Appl. Phys. A* **60**, 247 (1995).
4. Lindroos, M., and Bansil, A., *Phys. Rev. Lett* **77** 2985 (1996) and references therein.
5. Saini, N. L., Avila, A., Bianconi, A., Lanzara, A., Asensio, M. C., Tajima, S., Gu G. D., and Koshizuka, N., *Phys. Rev. Lett.* **79**, 3464 (1997).
6. Chuang, Y.-D., Gromko, A. D., Dessau, D.S., Aiura, Y., Yamaguch, Y., Oka, K., Arko, A.J., Joyce, J., Eisaki, H., Uchida, S.I., Nakamura K., and Ando, Y., *Phys. Rev. Lett.* **83**, 3717 (1999).
7. Feng, D. L., Zheng, W. J., Shen, K. M., Lu, D. H., Ronning, F., Shimoyama, J.-I., Kishio, K., Gu, G., Van der Marel, D., Shen, Z.-X., cond-mat/9908056.
8. Fretwell, H.M., Kaminski, A., Mesot, J., Campuzano, J.C., Norman, M.R., Randeria, M., Sato, T., Gatt, R., Takahashi, T., and Kadowaki, K., cond-mat/9910221.
9. Saini, N. L., Bianconi, A., Lanzara, A., Avila, A., Asensio, M. C., Tajima, S., Gu G. D., and Koshizuka, N., *Phys. Rev. Lett.* **82**, 3467 (1999); Saini, N. L., Bianconi, A., Lanzara, A., Avila, A., Asensio, M. C., Tajima, S., Gu G. D., and Koshizuka, N., *Physica* **C317-318**, 304 (1999).
10. Norman, M.R., Ding, H., Randeria, M., Campuzano, J.C., Yokoya, T., Takeuchi, T., Takahashi, T., Mochiku, T., Kadowaki, K., Guptasarma, P. and Hinks, D.G., *Nature* **392**, 157 (1998).; Ding, H., Yokoya, T., Campuzano, J.C., Takahashi, T., Randeria, M., Norman, M.R., Mochiku, T., Kadowaki, K., and Giapintzaki, J., *Nature* **382**, 51(1996).
11. Loeser, A.G., Shen, Z.X., Dessau, D.S., Marshall, D.S., Park, C.H., Fournier, P and Kapitulnik, A., *Science* **273**, 325 (1996); Loeser, A.G., Dessau, D.S., and Shen, Z.X., *Physica*C **263**,208 (1996).
12. Avila, J., Casado, C., Asensio, M. C., Munoz J. L., and Soria, F., *J. Vacuum Sci. Tech. A* **13**, 1501 (1995).
13. Randeria, M., Ding, H., Campuzano, J. C., Bellman, A., Jenning, G., Yokoyo, T., Takahashi, T., Katayama-Yoshida, H., Mochiku, T. and Kadowaki, K., *Phys. Rev. Lett.* **74**, 4951 (1995).
14. Ushio, H., and Kamimura, H., *J. Phys. Soc. Japan* **64**, 2585 (1995).
15. LaRosa, S., Kelley, R. J., Kendziora, C., Margaritondo, G.,. Onellion M.,and Chubukov, A., *Solid State Comm.* **104**, 459 (1997).
16. Andersen, O. K., Jepsen, O., Liechtenstein A. I., and Mazin, I. I., *Phys. Rev.* **B49**, 4145 (1994); Singh, D. J., and Pickett, W. E., *J. Superconductivity* **8**, 583 (1995).
17. Bansil, A., and Lindroos, M., *J. Phys. Chem. Solids* **59**, 1879 (1998).
18. Bianconi, A., Missori, M., *J. Phys. I* (France) **4**, 361 (1994).

19. Bianconi, A., Saini, N. L., Lanzara, A., Missori, M., Rossetti, T., Oyanagi, H., Yamaguchi, H., Oka, K., Ito, T., *Phys. Rev. Lett* **76**, 3412 (1996); Saini, N.L., Lanzara, A., Bianconi, A., Oyanagi, H., Yamaguchi, H., Oka, K., Ito, T., *Physica* C **268**, 121 (1996).
20. Zhou, J.-S., Bersuker, G.I., and Goodenough, J.B., *J. of Supercond* **8**, 541 (1995).
21. Mihailovich D., and Müller, K. A., *High T_c Superconductivity : Ten years after the Discovery* (Nato ASI Series-Applied Sciences, *Vol.* 343) edKaldis,. E., Liarokapis, E. and Müller, K. A., (Dordrecht, Kluwer) pag. 243-256.
22. Müller, K. A., Zhao, Guo-meng, Conder, K., and Keller, H., *J. Phys. Condens. Matter* **10**, L291-L296 (1998).
23. Ashkenazi, J., *Stripes and Related Phenomena*, edited by Bianconi A., and Saini N. L., (Kluwer Academics-Plenum Publisher) and references therein.
24. Bianconi, A., Saini, N. L., Rossetti, T., Lanzara, A., Perali, A., Missori, M., Oyanagi, H., Yamaguchi, H., Nishihara, Y., and Ha, D.H., *Phys. Rev. B* **54**, 12018 (1996); Bianconi, A., Lusignoli, M., Saini, N. L., Bordet, P., Kvick, Å., Radaelli, P.G., *Phys. Rev. B* **54**, 4310 (1996).
25. see also the special issue on 'Stripe Lattice Instabilities and High T_c Superconductivity' edited by Bianconi, A., and Saini, N. L., [*J. Supercond.* **10**, No. 4 (1997)].
26. see for example, Seino, Y., Kotani, A., and Bianconi, A., *Jour. Phys. Soc. of Japan* **59**, 815 (1990).
27. Bianconi, A., Valletta, A., Perali, A., and Saini, N. L., *Physica C***296**, 269 (1998).
28. Markiewicz, R.S. cond-mat/9911108.

The GW approximation: Theory and Application to YH_3

Takashi Miyake[*], Ferdi Aryasetiawan[*]
Hiori Kino[†] and Kiyoyuki Terakura[+]

[*]*Joint Research Center for Atom Technology,*
Angstrom Technology Partnership,
1-1-4 Higashi, Tsukuba, Ibaraki 305-0046, Japan
[†]*Institute for Solid State Physics, University of Tokyo,*
Roppongi, Minato-ku, Tokyo 106-8666, Japan
[+] *Joint Research Center for Atom Technology,*
National Institute for Advanced Interdisciplinary Research,
1-1-4 Higashi, Tsukuba, Ibaraki 305-8562, Japan

Abstract. *Ab initio* calculations of excited state properties of solids have become feasible with a steady increase in computing power. A suitable method for studying excited-state properties of extended systems is the Green function method which requires knowledge of the self-energy operator. A simple and fruitfull approximation to the self-energy beyond the Hartree-Fock approximation that takes into account screening is the GW approximation (GWA). It has been found to be successful in describing quasiparticle energies in a wide range of systems. Despite its success, there are some theoretical difficulties. The GWA has been found to be inadequate for describing satellite structures and self-consistent GW calculations tend to worsen the good agreement with experiment. Recent development beyond the GWA to improve the satellite description as well as the self-consistency issue will be discussed. As an application of the GWA we consider YH_3. Metal hydrides show a reversible metal-insulator transition in the visible range, making them attractive for optical switch. Density-functional calculations give incorrectly a metallic state unless we assume a complicated structure but GW calculations suggests that YH_3 is a normal insulator.

INTRODUCTION

A useful tool for studying the spectroscopic properties of materials is the Green function technique. To calculate the Green function knowledge of the self-energy is required. Various one-particle theories amount essentially to approximating the self-energy. In the Hartree-Fock approximation (HFA), for example, the self-energy is approximated by a bare exchange potential and in the Kohn-Sham density functional theory the local exchange-correlation potential may be regarded as an approximate self-energy although the interpretation is not clear.

An approximation to the self-energy which has been found to be successful in describing quasiparticle energies in a wide range of systems is the GW approximation [1–3]. Loosely speaking, the GWA is a generalization of the HFA by including the effects of screening. Thus, in the language of the Green function, the bare Coulomb interaction is replaced by a screened interaction.

Despite its success, the GWA has some difficulties. Notable among these is its rather poor description of satellite structures, even in sp systems where the GWA is expected to work best and the situation becomes worse when it comes to 3d or 4f systems. We discuss two approaches for improving the satellite description, the cumulant expansion method and the T-matrix theory.

Another fundamental problem is the issue of self-consistency. Recent works have shown that fully self-consistent calculations tend to worsen the good agreement with experiment obtained in one-iteration calculations. Self-consistency is evidently desirable since it eliminates the dependency on the initial starting Green's function. We believe the poor self-consistent results are due to the unphysical feature of the self-consistent scheme. Rather than abandoning the GWA which has been found to be so fruitful we propose a partial self-consistent scheme which retains the physical feature of the one-iteration GW.

As an application of the GWA we consider YH_3. Recent experiments on metal hydrides have found that these materials show a reversible metal-insulator transition which occurs in the visible range, making them attractive for application to optical switch. Density-functional calculations give incorrectly a metallic state unless we assume a complicated structure.

THEORY

The GW approximation

The time-ordered Green function is defined as [6]

$$iG(1,2) = \langle 0|T[\psi(1)\psi^\dagger(2)]|0\rangle \tag{1}$$

where we have used a short-hand notation $1 = (\mathbf{r}_1, t_1)$. In the sudden approximation, the photoemission spectrum may be related to the imaginary part of the Green function [4,5]

$$I(k,\omega) = 2\pi \sum_f |\langle f|A.p|0\rangle|^2 \delta(\omega - k^2/2 - \varepsilon_f)$$
$$\propto \text{Im } G(k,\omega) \tag{2}$$

The self-energy is defined by the following equation of motion of the Green function:

$$[i\partial_{t_1} - H_0(1) - \phi(1)]G(1,2)$$
$$- \int d3 \Sigma(1,3)G(3,2) = \delta(1-2) \tag{3}$$

$\phi(1)$ is an external probing field, used as a mathematical device to calculate the self-energy (Schwinger's functional derivative technique). The self-energy can be expressed in terms of response functions:

$$\Sigma(1,2) = -i \int d3d4 \, G(1,3)v(1-4)\frac{\delta G^{-1}(3,2)}{\delta\phi(4)} \tag{4}$$

From Eq. (3) we have

$$G^{-1} = i\partial_t - H_0 - \phi - \Sigma \tag{5}$$

$$\frac{\delta G^{-1}(1,2)}{\delta\phi(3)} = -\delta(1-2)\left[\delta(1-3) + \frac{\delta V_H(1)}{\delta\phi(3)}\right] - \frac{\delta\Sigma(1,2)}{\delta\phi(3)}$$

$$= -\delta(1-2)\epsilon^{-1}(1,3) - \frac{\delta\Sigma(1,2)}{\delta\phi(3)} \tag{6}$$

The GWA takes into account the response of the electrons within the Hartree approximation (Time-dependent Hartree or RPA), i.e., it takes into account the first term only in the above equation. With $W = \epsilon^{-1}v$ we then have after Fourier transformation

$$\Sigma(\mathbf{r},\mathbf{r}',E) = \frac{i}{2\pi}\int d\omega \, G(\mathbf{r},\mathbf{r}',E+\omega)W(\mathbf{r},\mathbf{r}',\omega) \tag{7}$$

Using a non-interacting Green's function, the real and imaginary parts of the self-energy can be calculated explicitly:

$$\mathrm{Re}\Sigma(\mathbf{r},\mathbf{r}',\omega)$$
$$= -\sum_{\mathbf{k}n}^{\mathrm{occ}} \psi_{\mathbf{k}n}(\mathbf{r})\psi_{\mathbf{k}n}^*(\mathbf{r'})\mathrm{Re}W(\mathbf{r},\mathbf{r}',\varepsilon_{\mathbf{k}n}-\omega) \tag{8}$$

$$+ \sum_{\mathbf{k}n} \psi_{\mathbf{k}n}(\mathbf{r})\psi_{\mathbf{k}n}^*(\mathbf{r'})W_{\mathrm{COH}}(\mathbf{r},\mathbf{r}',\omega-\varepsilon_{\mathbf{k}n}) \tag{9}$$

$$\mathrm{Im}\,\Sigma(\mathbf{r},\mathbf{r}',\omega \le \mu)$$
$$= \pi \sum_{\mathbf{k}n}^{\mathrm{occ}} \psi_{\mathbf{k}n}(\mathbf{r})\psi_{\mathbf{k}n}^*(\mathbf{r'})\,\mathrm{Im}\,W(\mathbf{r},\mathbf{r}',\varepsilon_{\mathbf{k}n}-\omega)\theta(\varepsilon_{\mathbf{k}n}-\omega)$$
$$\tag{10}$$

$$\mathrm{Im}\,\Sigma(\mathbf{r},\mathbf{r}',\omega > \mu)$$
$$= -\pi \sum_{\mathbf{k}n}^{\mathrm{unocc}} \psi_{\mathbf{k}n}(\mathbf{r})\psi_{\mathbf{k}n}^*(\mathbf{r'})\,\mathrm{Im}\,W(\mathbf{r},\mathbf{r}',\omega-\varepsilon_{\mathbf{k}n})\theta(\omega-\varepsilon_{\mathbf{k}n})$$
$$\tag{11}$$

$$W_{\text{COH}}(\mathbf{r},\mathbf{r}',\omega) \equiv \frac{1}{\pi}\text{P}\int_0^\infty d\omega' \frac{\text{Im}W(\mathbf{r},\mathbf{r}',\omega')}{\omega-\omega'}$$

$$W_{\text{COH}}(\mathbf{r},\mathbf{r}',0) = \frac{1}{2}V_{\text{scr}}(\mathbf{r},\mathbf{r}',0) \tag{12}$$

$$W = v + V_{\text{scr}} \tag{13}$$

where V_{scr} is a frequency-dependent screening potential defined by the above equation. The first term in Eq. (8) can be interpreted as a frequency-dependent screened exchange and the second term as the potential arising from the Coulomb hole (COH), i.e. the Coulomb potential arising from the screening charge [1,2].

The cumulant expansion

It has been found that the GWA does not give a satisfactory description of the satellite structure. We may distinguish two types of satellites, one originating from the long-range correlations resulting in plasmon-type of satellites and another of atomic character originating from short-range correlations resulting in smaller energy satellites a few eV below and above the main quasiparticle peak, which may sometimes be interpreted as the lower and upper Hubbard bands. The cumulant expansion [7–10,4] takes into account higher order coupling to the plasmon field and as such improves the description of the long-range satellites.

The cumulant expansion is based on the expansion of the Green function rather than the self-energy. We write the zeroth order and full Green function as

$$G_0(t) = ie^{-i\varepsilon t}$$

$$G(t) = ie^{-i\varepsilon t + C(t)}$$
$$= G_0(t)[1 + C(t) + \frac{1}{2}C^2(t) + \ldots] \tag{14}$$

The quantity $C(t)$ is defined to be the cumulant. From the Dyson equation we also have

$$G = G_0 + G_0\Sigma G_0 + G_0\Sigma G_0\Sigma G_0 + \ldots \tag{15}$$

Identifying C term by term we obtain

$$G_0 C = G_0 \Sigma G_0 \tag{16}$$

so that

$$C(t) = -i\int_0^t dt' \int_{t'}^\infty d\tau\, \Sigma(\tau) e^{i\varepsilon \tau} \tag{17}$$

This formal derivation hardly gives any justification for the use of the cumulant expansion. The justification comes from the consideration of a core electron coupled to a plasmon field:

$$H = \epsilon c^\dagger c + \omega_p b^\dagger b + g c c^\dagger (b^\dagger + b) \tag{18}$$

where c and b represent the core electron and the boson (plasmon) operator respectively, ω_p and g are the plasmon energy and the coupling constant respectively. The exact spectral function can be calculated analytically and using $\Sigma = iGW$ in the cumulant, in can be shown that the cumulant expansion yields the *exact* spectral function [8].

Applications of the cumulant expansion to Na and Al have successfully reproduced the multiple plasmon satellites seen in XPS measurements and remedied the problem encountered in the GWA [11].

The T-matrix theory

In many strongly correlated systems, one observes in a photoemission experiment the presence of a satellite structure a few eV below the main peak [12]. The origin of this satellite is due to the presence of two or more holes in a narrow band after a photoelectron is emitted. The correlation between these holes or particles is of short-range nature and it is not captured by the GWA which mainly takes into account the coupling of the electrons to the plasmons. A natural extension of the GWA is therefore to include short-range correlation which can be partially accounted for by the T-matrix theory [13–20] which describes multiple scattering between two holes or two particles.

Physically, the T-matrix describes multiple scattering between two holes or electrons. It is defined by the following Bethe-Salpeter equation [6]:

$$\begin{aligned}T_{\sigma_1\sigma_2}(1,2|3,4) &= U(1,2)\,\delta(1-3)\delta(2-4) \\ &+ U(1,2)\int d1'd2'\, K_{\sigma_1\sigma_2}(1,2|1',2')T_{\sigma_1\sigma_2}(1',2'|3,4)\end{aligned} \tag{19}$$

where U is a screened Coulomb interaction and K is a two-particle propagator. We have used a short-hand notation $1 \equiv (\mathbf{r}_1, t_1)$ and σ labels the spin.

The kernel K or the two-particle propagator is given by

$$K_{\sigma_1\sigma_2}(1,2|1',2') = iG_{\sigma_1}(1',1)G_{\sigma_2}(2',2) \tag{20}$$

where G_σ is a time-ordered single-particle Green function. The explicit form of the Fourier transformed kernel K is given by

$$K_{\sigma_1\sigma_2}(\mathbf{r}_1,\mathbf{r}_2|\mathbf{r}'_1,\mathbf{r}'_2;\omega) =$$

$$-\sum_{ij}^{occ} \frac{\psi_{i\sigma_1}(\mathbf{r'}_1)\psi_{i\sigma_1}^*(\mathbf{r}_1)\psi_{j\sigma_2}(\mathbf{r'}_2)\psi_{j\sigma_2}^*(\mathbf{r}_2)}{\omega - \varepsilon_{i\sigma_1} - \varepsilon_{j\sigma_2} - i\delta}$$
$$+\sum_{ij}^{unocc} \frac{\psi_{i\sigma_1}(\mathbf{r'}_1)\psi_{i\sigma_1}^*(\mathbf{r}_1)\psi_{j\sigma_2}(\mathbf{r'}_2)\psi_{j\sigma_2}^*(\mathbf{r}_2)}{\omega - \varepsilon_{i\sigma_1} - \varepsilon_{j\sigma_2} + i\delta} \quad (21)$$

where $\psi_{i\sigma}$ is a Bloch state. The first term on the right hand side is due to hole-hole scattering and the second to particle-particle scattering. This expression is similar to the RPA polarization propagator but the states are either both occupied or unoccupied.

The self-energy obtained from the T-matrix consists of a direct term

$$\Sigma_{\sigma_2}^d(4,2) = -i \sum_{\sigma_1} \int d1 d3 \, G_{\sigma_1}(1,3) T_{\sigma_1,\sigma_2}(1,2|3,4) \quad (22)$$

and an exchange term

$$\Sigma_{\sigma_2}^x(3,2) = i \int d1 d4 \, G_{\sigma_2}(1,4) T_{\sigma_2,\sigma_2}(1,2|3,4) \quad (23)$$

Using the spectral representations of G and T the self-energy can be written explicitly as

$$\text{Im } \Sigma_{\sigma_2}^d(\mathbf{r}_4, \mathbf{r}_2; \omega > \mu) =$$
$$\int d^3 r_1 d^3 r_3 \sum_{\mathbf{k'}n'\sigma_1}^{occ} \psi_{\mathbf{k'}n'\sigma_1}(\mathbf{r}_1) \psi_{\mathbf{k'}n'\sigma_1}^*(\mathbf{r}_3)$$
$$\times \text{Im } T_{\sigma_1\sigma_2}(\mathbf{r}_1, \mathbf{r}_2|\mathbf{r}_3, \mathbf{r}_4; \omega + \varepsilon_{\mathbf{k'}n'\sigma_1})$$
$$\times \theta(\omega + \varepsilon_{\mathbf{k'}n'\sigma_1} - 2\mu) \quad (24)$$

$$\text{Im } \Sigma_{\sigma_2}^d(\mathbf{r}_4, \mathbf{r}_2; \omega \leq \mu) =$$
$$-\int d^3 r_1 d^3 r_3 \sum_{\mathbf{k'}n'\sigma_1}^{unocc} \psi_{\mathbf{k'}n'\sigma_1}(\mathbf{r}_1) \psi_{\mathbf{k'}n'\sigma_1}^*(\mathbf{r}_3)$$
$$\times \text{Im } T_{\sigma_1\sigma_2}(\mathbf{r}_1, \mathbf{r}_2|\mathbf{r}_3, \mathbf{r}_4; \omega + \varepsilon_{\mathbf{k'}n'\sigma_1})$$
$$\times \theta(-\omega - \varepsilon_{\mathbf{k'}n'\sigma_1} + 2\mu) \quad (25)$$

The screened potential U is in principle frequency dependent but A static approximation is used in the present work. The theory is applied to Ce-α and the resulting spectrum is shown in Fig. (1). The positions of the satellites agree well with experiment [21].

SELF-CONSISTENCY

Recent self-consistent GW calculations on the electron gas [22] and Si [23] have produced both discouraging and encouraging results:

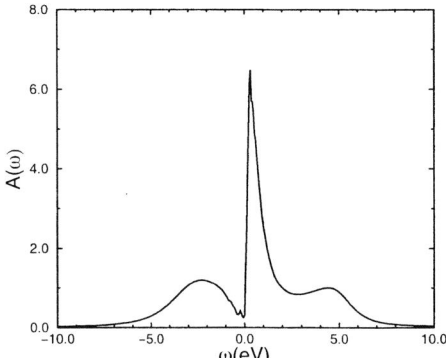

FIGURE 1. The calculated photoemission and inverse photoemission (BIS) spectra of Ce-α within the T-matrix approximation. The two satellites below and above the main quasiparticle peak correspond to the lower and upper Hubbard bands which are naturally missing in the LDA density of states.

- The electron gas bandwidth becomes larger than the free-electron one.
- The plasmon satellite almost disappears.
- The band gap in Si becomes larger than the experimental value (LDA= 0.53, one-iteration GW = 1.29, self-consistent GW = 1.91, exp=1.19 eV)
- The total energies for the electron gas are in very good agreement with the quantum Monte-Carlo result.

The worse results obtained from self-consistent GW appear to originate from the unphysical response function. In a one-iteration GW calculation, the polarization function $P = -iGG$ corresponds to the response of a non-interacting system whereas in a self-consistent calculation, $P = -iGG$ does not in general correspond to a response function of any system. Consequently, the physical picture of screening is lost and the screened potential W becomes an auxiliary rather than a physical quantity. As to why the self-consistent GW gives very good total energies for the electron gas, the reason is not yet clear.

Here, we propose simple a self-consistent scheme which ensures that the final result does not depend on the starting Green function and yet it retains the physical picture of the one-iteration GW. The scheme is illustrated in a diagram above. The right path corresponds to the fully self-consistent scheme and the left path to the proposed scheme, both indicated by thin arrows. A simplified form of this proposed self-consistent scheme was applied to NiO before [24] and is applied in this work to YH$_3$.

$$\boxed{H_0 \psi_k + V_{xc}\psi_k = \varepsilon_k \psi_k}$$

$$\Downarrow$$

$$\boxed{\psi_k, \varepsilon_k \to G = G_0}$$

$$\Downarrow$$

$$\boxed{P = -iGG}$$

$$\Downarrow$$

$$\boxed{\epsilon = 1 - vP \to W = \epsilon^{-1}v}$$

$$\uparrow \qquad \Downarrow \qquad \uparrow$$

$$\boxed{V_{xc} \leftarrow \text{Re}\,\Sigma} \leftarrow \boxed{\Sigma = iGW} \to \boxed{G^{-1} = G_0^{-1} - \Sigma}$$

$$\Downarrow$$

$$\boxed{E_k^{QP} = \varepsilon_k + \langle k|\Sigma - v_{xc}|k\rangle}$$

APPLICATION OF THE GWA TO YH$_3$

Recent experiments by Huiberts et al. [25] on yttrium and lanthanum hydrides (YH$_x$ and LaH$_x$) show that these materials undergo a metal-insulator transition at around $x = 2.8$ and turn transparent. At $x = 2$, they are metallic. Unlike other metal-insulator transitions, the transition in yttrium and lanthanum hydrides is reversible and it occurs in the visible range, thereby making them attractive for application to optical switch. Consequently these materials have drawn a lot of attention among both experimentalists and theorists. Subsequent experiments [26,27] have shown that the transition is common in a wide range of lanthanide hydrides [28]. There is also evidence that by choosing suitable rare-earth elements, the transition can be made very quick which is desirable in optical switch application.

The electronic structure of yttrium and lanthanide hydrides is unfortunately poorly understood. Theoretical works on these materials have followed along several lines. Band-structure calculations based on the local density approximation (LDA) of the density functional theory predict YH$_3$ and LaH$_3$ to be metals, with a band overlap of about 1 eV, instead of the experimentally observed semiconductors [29,30].

The discrepancy could be due to the structure effect. There may be a more complicated structure which is more stable and insulating. Kelly *et al.* made a study along this line. Starting from the HoD_3 structure, they found a broken symmetry structure, which is slightly more stable than the HoD_3 and has an LDA gap of 0.75 eV [31]. However, the structure is not consistent with neutron diffraction data [32,33]. Another difficulty is that the transition of lanthanum hydrides occurs without structural transformation: The system remains cubic during the transition [27]. This suggests that the transition is of electronic rather than structural origin.

Considering these observations, some theories, have been proposed that claim electron correlation at the hydrogen site is crucial to open a gap [34,35]. Ng *et al.* proposed an explanation based on a model Hamiltonian which takes into account the correlation effect on the hopping matrix element between hydrogen ions H_2^-. This model leads to narrowing of the valence hydrogen band causing it to separate from the conduction band resulting in a semiconducting gap [34]. Eder *et al.* also pointed out the importance of electron correlations at the hydrogen site [35]. They introduce a model Hamiltonian where the hopping between the hydrogens and the yttrium/lanthanides depends on the occupation number of the hydrogens. This leads to shift down of the valence band opening up a gap while keeping the hydrogen band wide. These arguments are, however, based on simple models. In the present work, we calculate the quasiparticle band structure with the *GW* approximation (GWA) [1,2]. We shall show that the self-energy correction is large enough to open a gap, though band-narrowing is not observed.

In Fig. 2(a), we show the band structures obtained by LDA and GWA. The valence band is predominantly hydrogen 1s with a mixture of Y (5s,5p and 4d) whereas the conduction band is dominated by the Y4d with some mixing of hydrogen 1s. The *GW* self-energy correction does not change the valence band so much. The self-energy correction is much larger for the conduction bands. It shifts up the bands by about 2 eV. Considering the rather poor starting Hamiltonian with a band overlap of almost 1 eV, this one-iteration result is very reasonable. Most *GW* calculations on real materials are performed with one iteration. We now apply a simplified form of the self-consistent scheme proposed in the previous section. We shifted the Y4d LMTO-orbital after the self-consistent LDA calculation, and did an extra LDA calculation for 1 iteration. Starting from this *shifted* Hamiltonian, we estimated the self-energy correction, and added the correction to the *unshifted* LDA eigenvalues. This procedure gives a quasiparticle band shown in Fig. 2(b). The overall picture is the same as Fig. 2(a). The valence band is unchanged, while the conduction band is raised up. The gap is larger than the unshifted one because of less screening, and is estimated to be 2.6 eV.

What happens then if we shift the Y4d band further? In Fig. 3, we plot the *GW* gap against the gap of the shifted starting Hamiltonian. (Here a negative gap means band overlap.) As we increase the shift, the *GW* gap increases monotonously. The self-energy correction, however, decreases, and the *GW* gap becomes smaller when

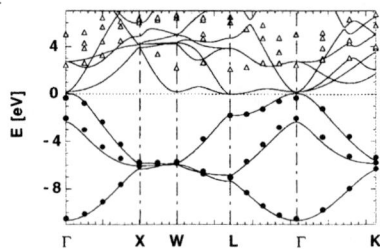

FIGURE 2. (a) The GW quasiparticle band structure of YH_3 with the BiF_3 structure. The filled circles (empty triangles) denote the valence (conduction) bands. The solid line represents the LDA band. (b) The same as (a) but the LDA Y4d bands are shifted up by 1.2 eV. The dotted horizontal line denotes the location of the Fermi level of the LDA band.

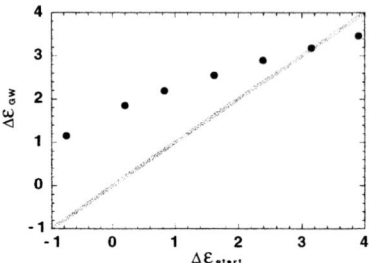

FIGURE 3. The GW gap ($\Delta\epsilon_{GW}$) against the gap of the starting Hamiltonian ($\Delta\epsilon_{start}$) in which Y4d band is shifted up compared to the LDA bands. The solid line is $\Delta\epsilon_{GW} = \Delta\epsilon_{start}$, plotted for comparison.

we start from a very large gap. The two gaps match when they are 3.2 eV. The experimental gap is 1.8 eV. (Note, however, that the structure used in the present calculation is different from the experimental one whose details are not known.)

Neither band-narrowing nor strong renormalization are observed. This suggests that correlations beyond LDA are not as strong as sometimes assumed. The main role of the self-energy correction is just raising up the conduction band, similar to semiconductors such as Si. This shift is large enough to remove the band overlap observed in the LDA. Therefore, the discrepancy between the LDA calculation and experiments can be explained in terms of electronic origin rather than of structural origin. The large discrepancy between the LDA and the experimental one could well be due to the poor performance of the LDA in band gap rather than to strong correlations. In this sense, our results suggest that YH_3 is a normal semiconductor like Si. Our results partially support the theoretical model proposed by Eder *et al.* but photoemission experiment is crucial in resolving the electronic structure of metal hydrides.

ACKNOWLEDGMENTS

This work is partly supported by the New Energy and Industrial Technology Development Organization (NEDO) and also by 'Research for the Future Program' of the Japan Society for the Promotion of Science.

REFERENCES

1. L.Hedin,*Phys.Rev.* **139**, A796 (1965).
2. L.Hedin and S.Lundquvist, in *Solid State Physics: Advances in Research and Applications*, edited by H.Ehrenreich, F.Seitz, and D.Turnbull (Academic, New York, 1969), Vol.23,p.1.
3. F. Aryasetiawan and O. Gunnarsson, *Rep. Prog. Phys.* **61**, 237 (1998).
4. C.-O. Almbladh and L. Hedin, *Handbook on Synchrotron Radiation* **1**, 686 ed. E. E. Koch (North-Holland, 1983).
5. L.Hedin, *J. Phys.: Condens. Matter* **11**, R489-528 (1999).
6. See, e.g., A. L. Fetter and J. D. Walecka, *Quantum Theory of Many-Particle Systems* (McGraw-Hill, New York, 1971).
7. P. Nozières and C. J. de Dominicis, *Phys. Rev.* **178**, 1097 (1969).
8. D. C. Langreth, *Phys. Rev. B***1**, 471 (1970).
9. B. Bergersen, F. W. Kus, and C. Blomberg, *Can. J. Phys.* **51**, 102 (1973).
10. L. Hedin, *Physica Scripta* **21**, 477 (1980).
11. F. Aryasetiawan, L. Hedin, and K. Karlsson, 1996 *Phys. Rev. Lett.* **77**, 2268 (1996).
12. S. Hüfner, *Photoelectron Spectroscopy*, Springer Series in Solid-State Sciences vol. 82 (Springer-Verlag, Berlin/Heidelberg 1996).
13. J. Kanamori, *Prog. Theor. Phys.* **30**, 275 (1963).
14. H. Suehiro, Y. Ousaka, and H. Yasuhara, *J. Phys. C: Solid State Phys.* **19**, 4247 (1986); *ibid.* **19** 4263 (1986).
15. H. Yasuhara, H. Suehiro, and Y. Ousaka, *J. Phys. C: Solid State Phys.* **21**, 4045 (1988).
16. D. R. Penn, *Phys. Rev. Lett.* **42**, 921 (1979).
17. A. Liebsch, *Phys. Rev. B* **23**, 5203 (1981).
18. C. Calandra and F. Manghi, *Phys. Rev. B* **45**, 5819 (1992).
19. Jun-ichi Igarashi, P. Unger, K. Hirai, and P. Fulde, *Phys. Rev. B* **49**, 16181 (1994).
20. M. Springer, F. Aryasetiawan, and K. Karlsson, *Phys. Rev. Lett.* **80**, 2389 (1998).
21. Y. Baer and W.-D. Schneider, in *Handbook on the Physics and Chemistry of Rare Earths, Vol 10*, ed. K. A. Gschneidner, Jr., L. Eyring, and S. Hüfner, (Elsevier, Amsterdam, 1987).
22. B. Holm and U. von Barth, *Phys. Rev. B* **57**, 2108 (1998).
23. W.-D. Schöne and A. G. Eguiluz, *Phys. Rev. Lett.* **81**, 1662 (1998).
24. F. Aryasetiawan and O. Gunnarsson, *Phys. Rev. Lett.* **74**, 3221 (1995).
25. J. N. Huiberts, R. Griessen, J. H. Rector, R. J. Wijngaarden, J. P. Dekker, D. G. de Groot, and N. J. Koeman, *Nature* **380**, 231 (1996).

26. J. N. Huiberts, R. Griessen, R. J. Wijngaarden, M. Kremers, and C. van Haesendonck, *Phys. Rev. Lett.* **79**, 3724 (1997).
27. M. Kremers, N. J. Koeman, R. Griessen, P. H. L. Notten, R. Tolboom, P. J. Kelly, and P. A. Duine, *Phys. Rev. B* **57**, 4943 (1998).
28. P. van der Sluis, M. Ouwerkerk, and P. A. Duine, *Appl. Phys. Lett.* **70**, 3356 (1997).
29. J. P. Dekker, J. van Ek, A. Lodder, and J. N. Huiberts, *J. Phys. Condens. Matter* **5**, 4805 (1993).
30. Y. Wang and M. Y. Chou, Phys. Rev. Lett. **71**, 1226 (1993); Y. Wang and M. Y. Chou, *Phys. Rev. B* **51**, 7500 (1995).
31. P. J. Kelly, J. P. Dekker, and R. Stumpf, *Phys. Rev. Lett.* **78**, 1315 (1997).
32. T. J. Udovic, Q. Huang, and J. J. Rush, *Phys. Rev. Lett.* **79**, 2920 (1997).
33. P. J. Kelly, J. P. Dekker, and R. Stumpf, *Phys. Rev. Lett.* **79**, 2921 (1997).
34. K. K. Ng, F. C. Zhang, V. I. Anisimov, and T. M. Rice, *Phys. Rev. Lett.* **78**, 1311 (1997).
35. R. Eder, H. F. Pen, and G. A. Sawatzky, *Phys. Rev. B* **56**, 10115 (1997).

Is a hole a single particle?

Changyoung Kim

Stanford Synchrotron Radiation Laboratory,[1]
Stanford, California 94309, USA

Abstract. Through an example of recent experimental evidence of spin-charge separation in one dimensional correlated electron systems, Sr_2CuO_3 and $SrCuO_2$, the seemingly natural concept of a 'hole' being a single particle is re-considered. It is argued that a hole as a particle should be an exception rather than the norm, contrary to one's usual perception. This implies that strange phenomena such as spin-charge separation in correlated systems may not be as strange as one may think.

I INTRODUCTION

Physics is a way of understanding nature by modeling. Therefore our understanding of nature can change depending on the model we use. Often a model based on another model is used to understand certain phenomena. In other words, previous knowledge is used to obtain new knowledge. This is, in fact, an extremely efficient way to better understand nature. Unfortunately, this leaves room for errors too. As models are based on models, derivative models fail if the base model turns out to be 'wrong'. Here 'wrong' may really mean wrong or may mean inapplicable. One historical example is the understanding of the solar system. Based on the model that the earth is the center of the motion, people tried to understand the mechanics of other planets. This of course did not work because the starting assumption is wrong. The prejudice prevented people from truly understanding the system. This happens quite often in today's scientific world and this will happen in the future. One way of reducing the possibility of such a mistake is knowing the limitations of the base model.

In condensed matter physics, some of the most important progresses come from band theory which is based on the non-interacting electron picture where the dominating terms are the kinetic and potential energies, and the electron correlation is ignored. This has had a tremendous contribution to the understanding of solids. Even though this has been used for more than half a century, it is still the dominating model in almost every sense. The reason is simple: it is relatively easy

---

[1] The Stanford Synchrotron Radiation Laboratory is operated by the U. S. DOE, office of Basic Energy Sciences, Division of Chemical Sciences.

and more importantly it works well for many kinds of solids. It and its extended concepts are used in theories of semi-conductors to conventional superconductors. The prevalence of it is so strong that much of our reasoning is usually based on this model, and the derivatives of the model are quite often accepted without much thinking. As the non-interacting electron picture can not be used where electron-electron correlation is very strong, the derivatives of it can not be used in that area. Unfortunately, this is not the case and derivatives are often used even when the non-interacting electron picture is not. Consequently, the misuse of the models greatly limits our reasoning.

With what is discussed above in mind, we wish to visit some fundamental concepts in solid state physics. The main subject considered here is a 'hole'. A hole was originally a spin off concept of the band picture and is widely used in solid state physics. It is quite often knowingly or unknowingly assumed to be a single particle and an equivalent of the positron. Through an example of spin-charge separation, it will be shown that a hole in some cases does not behave like a single particle in correlated systems. Rather, it may behave like a composite particle. By further examination of the concept of a hole, it is argued that a hole behaving like a single particle should be considered an exception and a strange phenomenon such as a hole showing spin-charge separation can be general behavior.

II REVISITING RELATED CONCEPTS

To proceed further, some of the concepts/definitions need to be clarified. First, the meaning of a 'single particle' should be considered. In dictionaries, it is defined as a very minute portion or quantity of matter, but the general concept of a single particle used in physics is somewhat different. It may be defined within the scope of the discussion here as *a state of which the wavefunction can be expressed in one variable for each physical quantity*. One should note that geometrical confinement is not required to be a single particle contrary to the dictionary definition. In addition, this definition allows collective modes to be single particles. Examples are electrons, phonons, magnons, plasmons, and holes.

'Being independent' or 'non-interacting' should also be considered. Physically, 'being independent' means objects do not affect each other. Quantum mechanically, it means that the total wave function can be expressed as a simple product form of individual object's wave function if we ignore the statistics. Therefore, the wave function of a non-interacting electron system can be expressed as a simple product of single electron wave functions (of course the real wave function is expressed as a single Slater determinant rather than a simple product form due to statistics. However, there is no difference between the two for the sake of the discussion here).

Lastly, we consider the meaning of a 'hole'. It is defined as a vacant orbital in a band [1]. The concept of a hole was introduced to describe an one electron removal state of a non-interacting electron *system*. Therefore, it is important to note that a hole actually refers to a state of an electron system, not just the missing electron.

Then how can a system of $N-1$ electrons be described as a single particle rather than $N-1$ particles? Figure 1 shows the ground state and one electron removal state of a non-interacting electron system. The *difference* between the two states is the single electron state that the missing electron used to occupy. Everything else is the same between the two states of the system. This allows the one electron removal state of the system (with $N-1$ electrons) to be described by a single variable (or a single particle) instead of $N-1$ which would be needed in general. This greatly simplifies the problem and helps to understand the physics intuitively. It should be stressed that this reduction of number of variables is possible only because the electrons are non-interacting. Once the correlation is significant the one electron removal states of the system can not be described by one variable in general. Reduction of variables is not strange and there are many examples. Phonon is one of them. A phonon which describes *excited* states of a *system* of atoms behaves like a single particle and we do not need many variables to describe the system. In analogy with the phonon case, a hole in a non-interacting electron system can be understood as an single particle *excitation* of an electron *system*. For certain electron correlations, it was shown by Landau that at least the low

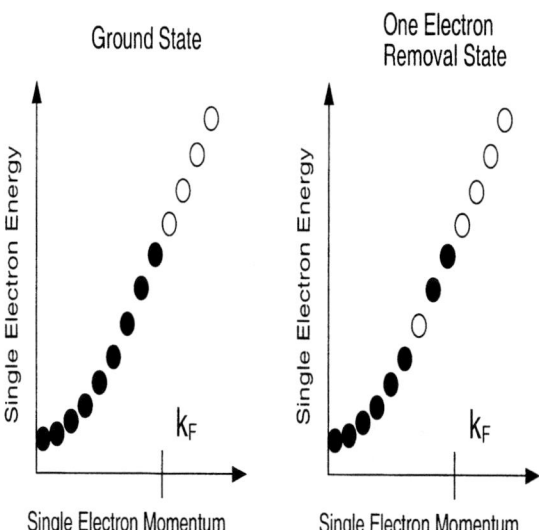

FIGURE 1. Illustration of a hole in a non-interacting electron system. The difference between the ground state and the one electron removal state of the non-interacting electron system is the missing electron in the single electron state. Therefore, the excitation from the ground state to the one electron removal state can be described by the missing single electron state or a hole.

energy part of the one electron removal states behave like a single particle (so called quasi-particle), which is now called Fermi liquid theory [2].

III SPIN-CHARGE SEPARATION

One natural question arises: is there a case where the one electron removal state does not behave like a single particle, even for low energy states? One such case is the spin-charge separation found in one dimensional correlated electron systems. This appears in the Tomonaga-Luttinger (TL) liquid [3] in one dimensional metal with correlated electrons and in the exact solution of the one dimensional Hubbard model by Lieb and Wu [4]. TL theory predicts the low energy excitations near the Fermi momentum can be factored into two independent excitations with spin and charge characters, respectively. Lieb and Wu also found that the excitations in the 1D Hubbard model show two independent excitations. The term spin-charge separation comes from the fact that the expected hole is replaced by two independent particles, called spinon and holon. Here spinon refers to the elementary spin excitation and holon to elementary charge excitation. It is also said to be the *decay* of a hole into a spinon and a holon.

The spin-charge separation can be redefined using the definitions discussed above. It now means that the one electron removal state of the system is expressed by the product of two single particle states instead of $N-1$, with the two independent single particles corresponding to the spin and charge excitations. Therefore, the seemingly exotic phenomenon is nothing but a special case where the system is described by two independent variables related to spin and charge excitations. With this given condition, it is not difficult to build the dispersion of the photoemission spectra which will be used as an experimental proof of spin-charge separation in one dimensional correlated system [5,6].

The photoemission process removes an electron from the system(N) (here system(N) refers to a system with N electrons) and ejects it into the vacuum [7]. This photoemitted electron carries information on the state of the system($N-1$) left behind through the energy and momentum conservation laws. Therefore, the photoemission closely mimics studies of the one electron removal states of the system [7]. Let us assume that the system($N-1$) (one electron removal state) can really be described by two independent variables relevant to spinon and holon. Energy and momentum conservation laws enforce the following relationships between the spinon, the holon and the photoelectron;

$$k + k_h - k_s = 0 \ (momentum\ conservation)$$
$$E + E_h - E_s = h\nu \ (energy\ conservation)$$

where k, k_s, k_h, E, E_s, and E_h are momenta and energies of the photoelectron, spinon and holon, respectively and $h\nu$ is the energy of the photon. The different signs on spinon and holon parts are due to the fact that photoemission process corresponds to *creation* of a holon and *annihilation* of a spinon [8].

Figures 2a and 2b illustrate the dispersion relations for holon and spinon, respectively, scaled in width by $2t$ and $\pi J/2$ as indicated by theoretical analysis [9] and experimental results [11]. For a spin chain with an AF interaction at half filling as in $SrCuO_2$ and Sr_2CuO_3, the holon band is empty while the spinon band is half filled and has a 'Fermi surface'. To create the lowest energy excitation, one can create the lowest energy holon at $k_h = 1$ and annihilate a spinon at the 'Fermi surface' with $k_s = 1.5$ (two circles in the figure). Then the momentum of the photohole becomes $k = -0.5$ (this corresponds to photoelectron momentum of $k_{\parallel} = 0.5$ in the figure) at which we observe the maximum of the bands in photoemission. Similar analysis leads to the expected picture in figure 2c. The reason that only one edge with holon character exists between 0.5 and 1 is that the spinon band is half filled.

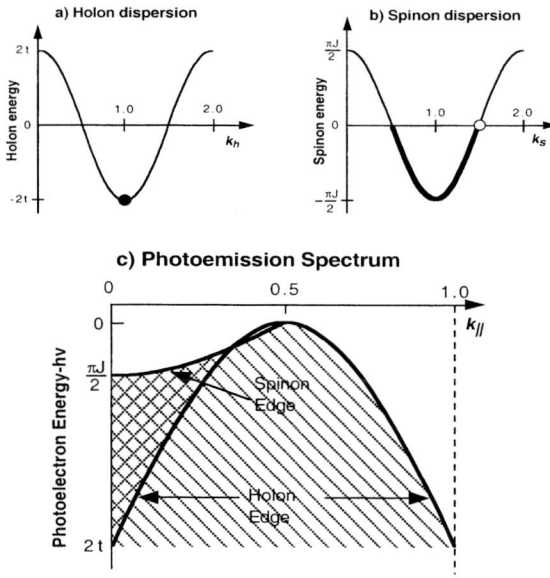

FIGURE 2. (a) and (b) Dispersions for holons and spinons. The holon band is empty while the spinon band is half filled and has the 'Fermi surface'. (c) The photoemission spectrum obtained from the two dispersions, and the energy and momentum conservation equations. In the region between 0 and 0.5, there are two edges in the spectra with the widths scaled by J and t. The edge with width $\pi J/2$ is due to spinon dispersion and the other with $2t$ is due to holon dispersion. Spectra in the shaded region show mixed excitations of spinons and holons. In the region between 0 and 0.5, strong spectral intensity is expected.

From 0 to 0.5, we have a heavily shaded region where strong photoemission signal is expected. This region is bounded by the spinon edge at lower excitation energy and holon edge at higher excitation energy.

Figure 3 shows the result of numerical calculation based on t-J model [5,12]. This can be regarded as a numerical experiment. The spin and charge correlation functions show that the spin and charge excitations are indeed independent in this model [5]. Note that the dispersion shown in figure 2 based on a very simple picture is in accord with the rigorous results presented in figure 3. The exact solution in this figure shows that the edges tend to have higher spectral intensities.

Sr_2CuO_3 and $SrCuO_2$ are Cu-O chain materials [13] and are ideal one dimensional correlated electron systems [13,14]. First experimental data supporting spin-charge separation was from $SrCuO_2$ [5,6] and results from more ideal Sr_2CuO_3 was reported later [15]. Figure 4 reproduces the most recent photoemission data from Sr_2CuO_3 reported by Fujisawa *et. al.* It is clear from the remarkable similarity between the theoretical picture in figure 2c and figure 3 and the dispersion shown in figure 4 that the spin-charge separation is the natural explanation of the experimental data from Sr_2CuO_3 which sharply contrast with anything one expects from the conventional band picture. That is, the description of the one electron removal states of one

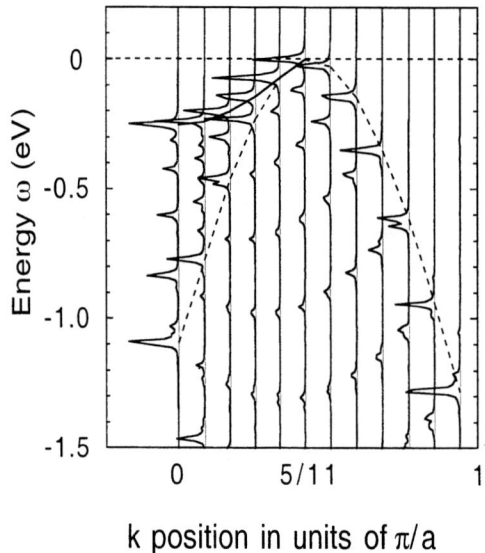

FIGURE 3. The calculated spectral function $A(k,\omega)$ in the t-J model with a ring of 22 sites. $J = 0.2$ eV and $t = 0.6$ eV were used. The energy ω is measured from the highest energy peak at $k = 5/11$. The edge with spinon(holon) character is marked with solid(dashed) line. Several peaks seen between the two edges rather than continuous spectra are due to a finite size effect.

dimensional correlated electron system indeed requires two variables.

IV CONCLUDING REMARKS

As shown above, description of the one electron removal states in one dimensional systems with correlated electrons requires two particles, spinon and holon. Then more questions arise: what happens in two or three dimensional strongly correlated systems? It is conceivable that a hole (the one electron removal state) may not be described by one or two variables. It may require $N-1$ variables after all as generally required. In that case, a hole behaves like $N-1$ particles rather than a single particle or a composite of two particles (spinon and holon, i.e., spin-charge separation). There are experimental indications that it may really be the case in high T_c superconductors, such as broad line shapes in photoemission [16]

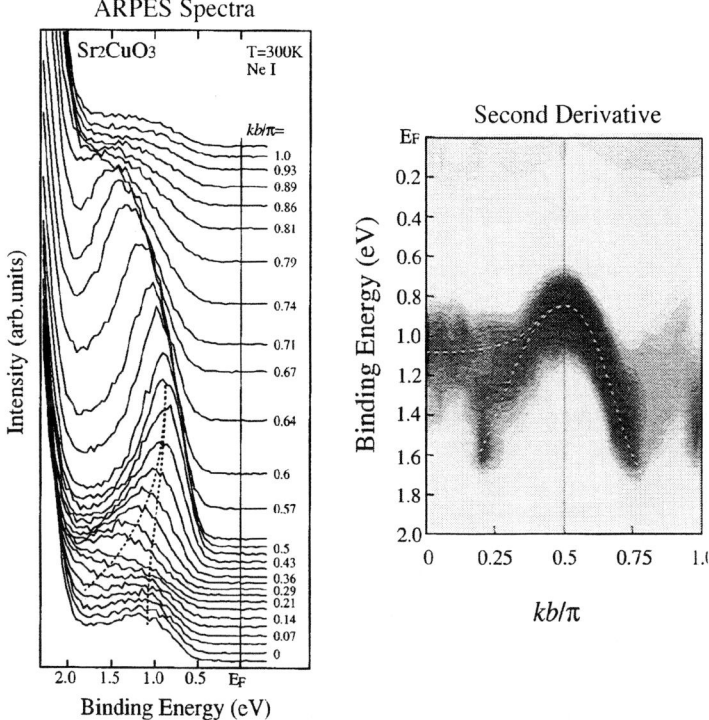

FIGURE 4. (a) ARPES data on S_2rCuO_3 from Ref. 15. The number on each spectrum shows the momentum parallel to the chain in units of π/b where b is the Cu-to-Cu distance along the chain. (b) The density plot of the second derivatives of the data. This plot shows the E vs. k relationship. The lines are guide to the eye for the peak positions. Note the good agreement between the density plot in this figure and the calculated spectra shown in figure 3.

and optical [17] spectra. This appears to be the case even for very low energy excitations [18]. This sounds very unusual and appears to be abnormal. However, it is stressed again that naturally we need as many variables as the number of particles to describe the system. Rather a hole ($N-1$ electron system) behaving like a single particle should be considered an exception. Fortunately, a hole behaves like a single particle for many systems such as semi-conductors and this greatly has helped our understanding of solids. But we should not let our prejudice of a hole being a single particle limit our imagination in correlated systems.

REFERENCES

1. C. Kittel, *Introduction to Solid State Physics*, (John Wiley & Sons, Inc., 6th Edition, New York, 1986).
2. D. Pines and P. Nozieres, *The Theory of Quantum Liquids*, (W. A. Benjamin, Inc., Amsterdam, 1966).
3. G. D. Mahan, *Many-particle physics* (Plenum Press, New York, 1990).
4. E. H. Lieb, and F. Y. Wu, Phys. Rev. Lett. **20**, 1445 (1968).
5. C. Kim, A. Y. Matsuura, Z.-X. Shen, N. Motoyama, H. Eisaki, S. Uchida, T. Tohyama, and S. Maekawa, Phys. Rev. Lett. **77**, 4054 (1996).
6. C. Kim, Z.-X. Shen, N. Motoyama, H. Eisaki, S. Uchida, T. Tohyama, and S. Maekawa, Phys. Rev. B, **56**, 15589 (1998).
7. S. Hufner, *Photoelectron spectroscopy : principles and application*, (Springer-Verlag, New York, 1995).
8. To be exact, an electron should be regarded as spinless charge and a spinon. Therefore, a hole consists of a holon and an antispinon (or spinon hole). This is the reason why the creation of a hole corresponds to creation of a holon and annihilation of a spinon.
9. In the limit of $J/t \to 0$, the Hamiltonian is decoupled into the spin and charge parts (see Ref. 10), and the spin part is mapped onto a system of interacting spinless fermions after the Jordan-Wigner transformation.
10. M. Ogata and H. Shiba, Phys. Rev. B **41**, 2326 (1990); S. Sorella and A. Parola, J. Phys.: Condens. Matter **4**, 3589 (1992).
11. H. Suzuura, H. Yasuhara, A. Furusaki, N. Nagaosa, and Y. Tokura, Phys. Rev. Lett. **76**, 2579 (1996); H. Yasuhara *et al.*, unpublished.
12. T. Tohyama and S. Maekawa, J. Phys. Soc. Jpn. **65**, 1902 (1996).
13. N. Motoyama, H. Eisaki, and S. Uchida, Phys. Rev. Lett. **76**, 3212 (1996).
14. M. Takigawa *et al.*, (unpublished).
15. H. Fujisawa, *et al*, Phys. Rev. B **59**, 7358 (1999).
16. Z.-X. Shen and D. S. Dessau, Phys. Rep. **253**, 1 (1995).
17. S. Uchida, *et al.*, Phys. Rev. B **43**, 7942 (1991).
18. C. Kim, *Superconducting and Related Oxides: Physics and Nanoengineering III*, eds. D. Pavuna and I. Bozovic, SPIE Proc. 3481 (SPIE, Bellingham, p. 17(1998)).

Calculation and Interpretation of X-ray Spectroscopies with Green's Function Multiple Scattering Theory

A. L. Ankudinov, A. Nesvishskii, and J. J. Rehr

Dept. of Physics, University of Washington, Seattle, WA 98195-1560

Abstract.
All theories of x-ray based spectroscopies are based fundamentally on quantum electrodynamics (QED), which fully describes the interaction between the electromagnetic field and matter. Since there is no exact solution for the many-body QED Hamiltonian, approximations specific for each particular spectroscopy have been developed separately. Thus for Extended X-ray Absorption Fine Structure (EXAFS) the scattering potential is well approximated by a superposition of individual atomic electron densities. However this can be a bad approximation near absorption edges, where effects of charge transfer and non spherical potentials can be important. In an attempt to develop a general treatment of x-ray spectroscopies, we have developed an all electron, relativistic, self-consistent real space Green's function (RSGF) code FEFF8, for multiple-scattering calculations of x-ray absorption, emission, diffraction, etc. Simultaneous, auxiliary calculations of the projected density of states (DOS) are also carried out, which permit an interpretation of the spectra in terms of electronic structure. An illustrative example for Cu anomalous x-ray scattering is also presented, which emphasizes the importance of self-consistency for calculations and interpretation of near-edge data.

INTRODUCTION

The interaction between the electromagnetic field and matter is described fundamentally by quantum electrodynamics (QED), which yields a Hamiltonian in terms of the electrons and atomic nuclei. Since the laws of interaction are known, this Hamiltonian can be written out immediately for any chemical or solid state physics application. However this Hamiltonian can be solved exactly only for the system of two particles (e.g., the hydrogen atom), where it leads to spectacular agreement with experiment. However, there is no algebraic solution even for the helium atom, and one has to apply numerical methods to solve the general QED Hamiltonian.

I WAVE FUNCTION VS GREEN'S FUNCTION

There are two basic approaches to treat this problem numerically: wave function and Green's function methods. In the first, one seeks a many body wave function for ground and excited states of the system as function of particle coordinates. However, one can only handle a finite number of particles exactly without further approximations. In particular, one simply cannot store the wave function as a function of all particle coordinates for a system of more than ten particles. Thus atomic calculations are usually done within the independent electron approximation: i.e. the wave function is given by a Slater determinant or sometimes include configuration interactions, i.e. the wave function is linear combination of Slater determinants.

The second approach is based on Green's functions. Since the Coulomb interaction is two-particle, one needs only one- and two-particle Greens functions. Their knowledge is enough to calculate most physical properties of the system. A formal solution for single particle Green's function is given by the Hedin equations [1], and once known, the Bethe-Salpeter equation can be applied to find two-particle Green's function. As with the wave function approach, the Hedin equations must be solved numerically. The Green's function approach has been very successful in studies of the uniform electron gas, and leads to a formulation of density functional theory for electronic structure.

One of most useful approximations in solving these equations is the independent electron approximation. Within this approximation the one-body potential for a given electron configuration is known, and the problem reduces to a calculation of the wave function or Green's function for the Schrödinger (or Dirac) equation.

II MULTIPLE SCATTERING THEORY

Multiple scattering theory (MST) can be used to calculate both of these quantities. MST is based on separation of space into cells, within which the scattering can be treated to all orders in terms of the scattering t-matrix. Once the t-matrix is calculated, MST provides a way of combining individual cell solutions into a total Green's function or wave function via the Dyson or Lippman-Schwinger (LS) equations, respectively. The first electronic structure method based on the MS LS equation was formulated by Koringa [2], and developed independently by Kohn and Rostocker, and now is known as Koringa-Kohn-Rostocker (KKR) bandstructure method. A similar approach applied to finite clusters or molecules is the X_α-scattered wave (SW) method, of Johnson [3]. In the real space Green's function (RSGF) approach, originally proposed by Beeby [4] for disordered systems, one calculates directly the Green's function and not wave function by MST. A key advantage is that electron density is linearly proportional to the Green's function (more precisely to its imaginary part), instead of the square of the wave function. Thus the Green's function can be a more convenient object to manipulate for various sorts of averaging in disordered systems. For example, the coherent potential approximation (CPA) has been widely applied to study substitutional alloys. Many of the manipulations on the MS series are similar in the RSGF and SW methods. The main difference appears for bound states, where for an eigenenergy the MS series diverges. In the SW method one uses this fact to construct the secular equation (the KKR equations), while in RSGF one can avoid singular points by performing all calculations in the complex energy plane.

There have been many successful applications of MST calculations for various spectroscopies since the pioneering work of Vedrinskii et al. in 1974 [5], which gave x-ray absorption near edge spectroscopy (XANES) calculations for the L_3 edge of the SF_6 molecule in remarkable agreement with experiment. These calculations have been extended to many other spectroscopies, and few of the most recent examples include anomalous scattering [6], x-ray natural circular dichroism [7], and x-ray magnetic circular dichroism [8]. One of the main advantages of the MST formalism, is a geometrical interpretation of the x-ray spectra. For example, EXAFS is best interpreted in terms of scattering paths. Even in the XANES region, the peaks often exhibit strong sensitivity to local structure. However in the XANES region the complimentary electronic interpretation of the spectra in terms of local projected densities of states (LDOS) may be more appropriate.

Calculations of x-ray absorption can easily be generalized to calculations of the anomalous scattering amplitude f. The imaginary part of the anomalous scattering amplitude f'' is proportional to the absorption coefficient, and it's real part f' is formally connected to the imaginary part by a Kramers-Kronig transform. Calculations of the Thomson contribution f_0 to the x-ray scattering amplitude $|f|^2$, ($f = f_0 + f' + if''$) are also necessary to calculate the intensity of the Bragg peaks [9], but they are straightforward once the electron density is known. The real and imaginary parts of the Green's function, as calculated in RSGF, also satisfy a Kramers-Kronig relationship, and therefore one can simultaneously get the anomalous scattering amplitude and absorption coefficient in terms of the complex Green's function [10]. The energy dependence of Bragg peak intensities can be used to extract information about orbital ordering, local distortions around particular sort of atoms etc. X-ray scattering experiments have an advantage over absorption, since one can exploit the dependence on the initial and final polarization vectors. Thus measuring the energy dependence of different Bragg peaks allows one to separate the signal from different sites within the unit cell even for the same species of atom. In absorption this is impossible, since the relative contributions will be simply proportional to the number of atoms.

III CALCULATION OF ANOMALOUS X-RAY SCATTERING

We now illustrate the RSGF theory with our *ab initio*, self consistent x-ray spectroscopy code FEFF8 [11], which we have recently generalized for calculations of x-ray emission and anomalous x-ray scattering (AXS). Usually the RSGF approach calculates the retarded Green's function, and we found that for accurate calculations of AXS near the absorption edge, it is important to subtract the contribution to the Green's function from states below Fermi level. The results of the FEFF8 calculation of f' for the Cu K-edge are given in Fig. 1, and compared with the experimentally extracted scattering amplitude obtained with a differential KK transform [12].

These calculations were shifted slightly in energy to correct for the few eV error in the FEFF8 estimate of the absolute excitation edge, however, no shift or factor was used for the overall amplitude. Instead, the nearly constant contribution from all other edges was calculated, and a total energy shift appropriate for Cromer Liberman calculations [9] was added. The FEFF8 potentials were obtained self consistently, and no fitting parameters were used in the calculations. These results show that the subtraction of contributions below Fermi level is essential to obtain the edge cusp at the right position and amplitude. In diffraction experiments one often tries to avoid the region just above the edge, because the fine structure due to solid states effects makes XRD data analysis less reliable. Our calculations for Cu show remarkable agreement with experiment and suggest that one can actually use this fine structure to improve the determination of atomic coordinates within unit cell from the positions and intensities of Bragg peaks in XRD.

In summary, MST based calculations can be used to interpret x-ray spectra in

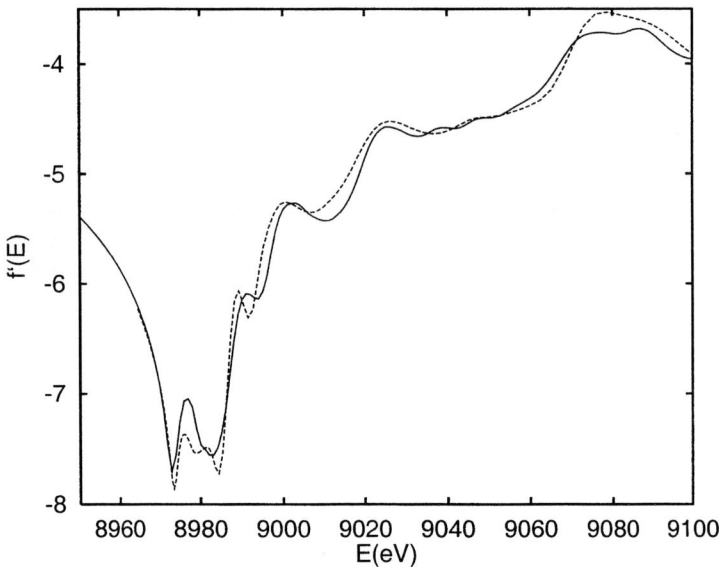

FIGURE 1. Anomalous scattering amplitude (f') for Cu K edge calculated with FEFF8 (dashes) vs experiment (solid).

terms of geometric and electronic structure. We have developed an *ab initio* relativistic, self-consistent code FEFF8, based on the MST for calculations of various x-ray spectroscopies using a real-space Green's function (RSGF) formalism. An illustrative application to solid state effects on anomalous x-ray scattering amplitudes is in semi-quantitative agreement with experiment. Although the goal of highly accurate calculations of XANES region remains challenging, further progress can lead to the development of more powerful analysis tools to better interpret x-ray spectra.

IV ACKNOWLEDGMENTS

This work was supported in part by US DOE grant DE-FG03-97ER45623.

REFERENCES

1. L. Hedin, S. Lundquist, Sol. St. Phys. **23**, 1 (1969).
2. J. Koringa, Physica **13**, 392 (1947).
3. K. H. Johnson, J. Chem. Phys. **45**, 3085 (1966).
4. J. L. Beeby, Proc. Roy. Soc. A, **302**, 113 (1967).
5. R. V. Vedrinskii, A. P. Kovtun, V. V. Kolesnikov, Y. F. Migal, E. V. Polozhentsev, and V. P. Sachenko, Izv. Akad. Nauk SSSR, Ser. Fizicheskaya, **38**, 8 (1974).
6. M. Benfatto, Y. Joly, and C. R. Natoli, Phys. Rev. Lett. **83**, 636 (1999).
7. C. R. Natoli, Ch. Brouder, Ph. Sainctavit, J. Goulon, Ch. Goulon-Ginet, and A. Rogalev, Eur. Phys. J. B 4, 1 (1998); J. Goulon, Ch. Goulon-Ginet,A. Rogalev, V. Gotte, C. Malgrange, Ch. Brouder, and C. R. Natoli, J. Chem. Phys. **108**, 6394 (1998).
8. A. L. Ankudinov, J. J. Rehr, Phys. Rev. B **52** 10 214 (1995).
9. D. T. Cromer, and D. Liberman, J. Chem. Phys. **53**, 1891 (1970).
10. R. V. Vedrinskii, v. L. Kraizman, A. A. Novakovich, et al., J. Phys. Cond. Mat. 4, 6155 (1992).
11. A. L. Ankudinov, B. Ravel, J. J. Rehr, and S. D. Conradson, Phys. Rev. B **58**, 7565 (1998).
12. J. O. Cross, M. Newville, J. J. Rehr, L. B. Sorensen, C. E. Bouldin, G. Watson, T. Grouder, G. H. Lander, and M. I. Bell, Phys. Rev. B **58**, 11215 (1998); J. O. Cross, M. Newville, private communication.

Treatment of non-collinear spin-structures in photo emission and X-ray absorption

H. Ebert*, J. Minár*, V. Popescu*, L. Sandratskii[†] and A. Mavromaras[†]

*Institut für Physikalische Chemie, Universität München, Butenandtstr. 5-13, D-81377 München, Germany

[†]Inst. für Festkörperphysik, TH Darmstadt, Hochschulstr. 2, D-64289 Darmstadt, Germany

Abstract. Many magnetic compounds possess a non-collinear spin-structure, i.e. the atomic magnetic moments may be oriented with respect to each other in an arbitrary way. When dealing with the electronic structure of such systems this situation can be accounted for in a straightforward way by using the relativistic multiple scattering theory for spin-polarised systems (SPR-KKR). The corresponding formalism is briefly outlined and its application to the studies of the photo emission and X-ray absorption is described. The results of calculations for UPdSn and hcp-Gd are presented.

INTRODUCTION

Most of the theoretical investigations of magnetic systems assume a collinear spin configuration, i.e. the individual atomic magnetic moments are assumed to be parallel or antiparallel to each other. Non-collinear spin structures are however very common. An important example for a non-collinear magnetic configuration is a magnetic solid at finite temperature: the atomic moments fluctuate and therefore deviate from the collinear directions. Another important example is the canting of the spin magnetic moments due to lattice imperfections or at a rough interface in a multilayer system [1]. There exist also numerous ordered non-collinear spin structures [2]. An example for an ordered non-collinear magnetic structure is the transition metal compound γ-FeMn [3]. Here the magnetic unit cell is not much larger than the chemical unit cell. On the other hand the magnetic spiral structures can possess very large or even infinite magnetic unit cells. Since the variation of the magnetic moments within the magnetic unit cell of the spiral structure is governed by a simple rule, the system possesses a generalized translational symmetry. In the case the spin-orbit coupling is neglected this symmetry can be used to essentially simplify the calculation for spiral structures [2].

Non-collinearity of the magnetic structure may have important consequences for the physical properties of a magnetic material. For example γ-FeMn and related compounds have been investigated as possible materials supplying the necessary pinning field for spin valve devices [4]. Recently, it was suggested that non-collinearity of the moments plays a crucial role for the invar properties of $Fe_{65}Ni_{35}$ [5]. Non-collinearity of the magnetic configuration has also very strong influence on the galvano-magnetic properties (GMR and TMR) of layered systems.

In the following a very flexible and efficient scheme is outlined that allows to deal with magnetic dichroism in various types of the electronic spectroscopy of material with non-collinear spin structure. This is achieved by describing the electronic structure by means of spin-polarized version of relativistic multiple scattering theory. Applications are presented for the valence band photo emission from UPdSn and the X-ray absorption spectra of hcp-Gd. In the case of UPdSn, the calculations are carried out for the experimental complex non-collinear magnetic structure. For hcp-Gd, results of model calculations for assumed spiral magnetic configurations are reported.

THEORETICAL FRAMEWORK

Magnetic dichroism is a consequence of the simultaneous occurrence of magnetic ordering and the spin-orbit coupling. Accordingly, the most satisfactory description of the magnetic dichroic phenomena is achieved by dealing with the electronic structure of the investigated system on the basis of the four-component Dirac formalism:

$$\mathcal{H}_D = \frac{c}{i}\vec{\alpha} \cdot \vec{\nabla} + \frac{e}{c}\vec{\alpha} \cdot \vec{A}(\vec{r}) + \frac{c^2}{2}(\beta - I) + V(\vec{r}) \,. \qquad (1)$$

Here all quantities have their usual meanings [6]. In particular, α_i ($i = 1, 2, 3$) and β are the standard Dirac matrices.

If exchange and correlation are accounted for within the framework of density functional theory, the scalar and vector potentials, $V(\vec{r})$ and $\vec{A}(\vec{r})$, respectively, will in general depend on the electronic current density (current density functional theory – CDFT). However, this most sophisticated density functional scheme still needs further development to allow its routine application to real systems. Fortunately, one may assume in most cases that the potential terms in Eq. (1) are primarily determined by the electronic spin magnetization $\vec{m}(\vec{r})$. This assumption leads to the relativistic version of the spin density functional theory (SDFT) that has been developed by MacDonald and Vosko [7] and Ramana and Rajagopal [8]. Within this approach the potential $V(\vec{r})$ in Eq. (1) consists of a spin-independent part that stems from the Coulomb potential $V_C(\vec{r})$ and the spin-averaged part $\overline{V}_{xc}(\vec{r})$ of the exchange correlation potential.

In addition one has a spin-dependent contribution for $V(\vec{r})$:

$$V_{\text{spin}}(\vec{r}) = \beta\vec{\sigma} \cdot \frac{\partial E_{\text{xc}}}{\partial \vec{m}(\vec{r})} = \beta\vec{\sigma} \cdot \vec{B}_{\text{xc}}(\vec{r}) \,, \tag{2}$$

that is determined by the dependence of the exchange-correlation energy E_{xc} on the spin magnetization $\vec{m}(\vec{r})$ and can be expressed by an effective exchange-correlation field $\vec{B}_{\text{xc}}(\vec{r})$.

Here it should be noted that the spin-orbit coupling is implicitly taken into account by working with the relativistic Dirac equation. Therefore the use of a potential that depends only on the spin magnetization $\vec{m}(\vec{r})$ will lead for magnetic systems to a finite orbital magnetization. What is missing within the SDFT scheme compared with the CDFT scheme is a potential term that feeds back the orbital current induced by the spin-orbit coupling current into the underlying Hamiltonian.

So far there are no geometrical restrictions connected with the potential term in Eq. (2). This means in particular that the effective exchange-correlation field $\vec{B}_{\text{xc}}(\vec{r})$ may vary from point to point not only in magnitude but also in orientation. This most general case has been studied for solids by Nordström and Singh [9] and for open shell atoms by Eschrig and Servedio [10]. Fortunately, for most applications it is sufficient to assume that $\vec{B}_{\text{xc}}(\vec{r})$ has a uniform orientation within each atomic sphere although this orientation is different for different atoms. In addition, for solids it is usually well justified to assume that $V(\vec{r})$ and $\vec{B}_{\text{xc}}(\vec{r})$ are spherically symmetric.

¿From Eq. (2) it is obvious that $V_{\text{spin}}(r)$ is represented by a 4×4-matrix. For the case that the magnetization is parallel to the z-axis the potential is a diagonal matrix with the elements specified by the spin-dependent potential terms $(V_{\text{xc}}^{\uparrow(\downarrow)}(r) - \overline{V}_{\text{xc}}(r))$. If the magnetization deviates from the z-direction, one has instead (with the argument r omitted):

$$V_{\text{spin}} = \beta R(\theta,\phi)^\dagger \begin{pmatrix} V_{\text{xc}}^\uparrow - \overline{V}_{\text{xc}} & & & \\ & V_{\text{xc}}^\downarrow - \overline{V}_{\text{xc}} & & \\ & & V_{\text{xc}}^\uparrow - \overline{V}_{\text{xc}} & \\ & & & V_{\text{xc}}^\downarrow - \overline{V}_{\text{xc}} \end{pmatrix} R(\theta,\phi) \,. \tag{3}$$

Here the terms $V_{\text{xc}}^{\uparrow(\downarrow)}(r)$ refer to the local reference system with the z-axis parallel to the spin moment m. The orientation of the local system is specified by the angles θ and ϕ (see Fig. 1). Finally, the transformation $R(\theta,\phi)$ transforms the potential matrix from the local to the global reference system [2].

The angles (θ_q, ϕ_q) must be specified for each atom q in the magnetic unit cell. For spin spirals the spin configuration is specified by a spiral vector $\vec{q}_s$:

$$\vec{m}_{nq} = m_q(\cos(\vec{q}_s \cdot \vec{R}_n + \phi_q)\sin\theta_q, \sin(\vec{q}_s \cdot \vec{R}_n + \phi_q)\sin\theta_q, \cos\theta_q) \,, \tag{4}$$

where $\vec{R}_n$ is a lattice vector. This situation is demonstrated in Fig. 1 for the case of $\vec{q}_s$ parallel to the global z-axis.

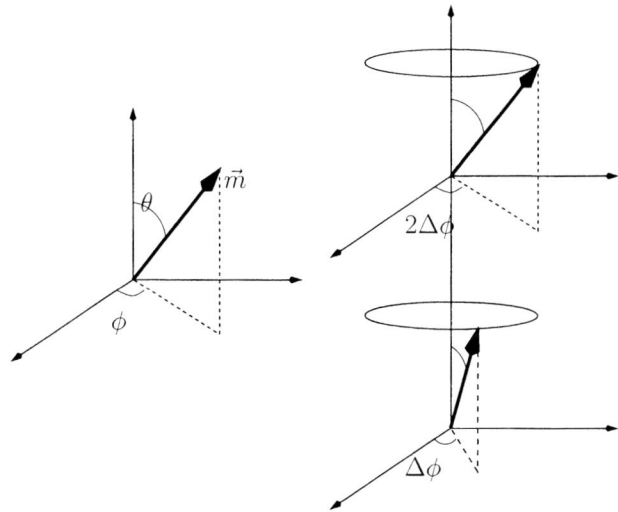

FIGURE 1. Left: Definition of the angles θ and ϕ that specify the orientation of the magnetization within an atomic cell with respect to the crystallographic axes. Right: Magnetic configuration for spin spirals. The angle ϕ is modified by $\Delta\phi$ from one atomic layer to a neighboring one.

In the following applications the relativistic spin-polarized version of multiple scattering theory or, equivalently, of the KKR method has been used. This scheme implies a decomposition of a band structure calculation into solving the single site scattering problem and a subsequent solution of the multiple scattering problem. The first step could be done straightforwardly using the potential term as given in Eq. (2). However, solving the single site Dirac equation is numerically much simpler adopting the local frame of reference for which $V_{\text{spin}}(r) = \beta\sigma_z B_{\text{xc}}(r)$ holds. The resulting single site t-matrices $\underline{t}^{q\,\text{loc}}$ for various sites q are then transformed to the global system according to [11]:

$$\underline{t}^{q\,\text{gl}} = \underline{R}(\theta_q, \phi_q)\, \underline{t}^{q\,\text{loc}}\, \underline{R}^\dagger(\theta_q, \phi_q)\,, \tag{5}$$

with the underline indicating a matrix with respect to the spin angular character $\Lambda = (\kappa, \mu)$ [6]. This allows one to deal with the multiple scattering problem by using the global frame of reference and inverting the corresponding KKR matrix

$$\underline{\underline{\tau}}^{\text{gl}} = \left[\left(\underline{\underline{t}}^{\text{gl}}\right)^{-1} - \underline{\underline{G}}\right]^{-1}. \tag{6}$$

Here the double underline indicates matrices with respect to the sites q and spin angular character Λ and $\underline{\underline{G}}$ is the free electron propagation or KKR structure constant matrix. Eq. (6) is either dealt with by direct inversion for a finite cluster of atoms or by means of a Fourier transformation in the case of a periodic system.

To calculate local electronic properties the site diagonal scattering path operator matrix $\tau^{qq\,\mathrm{gl}}$ is transformed back to the local system according to

$$\underline{\tau}^{qq\,\mathrm{loc}} = \underline{R}^\dagger(\theta_q, \phi_q)\,\underline{\tau}^{qq\,\mathrm{gl}}\,\underline{R}(\theta_q, \phi_q)\,. \tag{7}$$

With $\tau^{qq\,\mathrm{gl}}$ available the site projected Green's function is given by the standard expression

$$G^+(\vec{r}_q, \vec{r}_{q'}, E) = \sum_{\Lambda\Lambda'} Z_\Lambda^q(\vec{r}_q, E)\tau_{\Lambda\Lambda'}^{qq'\,\mathrm{loc}}(E)Z_{\Lambda'}^{q\times}(\vec{r}_{q'}, E)$$
$$- \sum_\Lambda Z_\Lambda^q(\vec{r}_<, E)J_\Lambda^{q\times}(\vec{r}_>, E)\delta_{qq'}\,, \tag{8}$$

with Z^q and J^q the regular and irregular solutions to the Dirac equation for site q.

Using the representation for the Green's function as given by Eq. (8) implies that all spectroscopic matrix elements $M^{\vec{q}_\mathrm{p}\lambda}$ are also evaluated in the local frame of reference. The corresponding relativistic electron-photon interaction operator $X^{\vec{q}_\mathrm{p}\lambda}$ is given by [18]:

$$X^{\vec{q}_\mathrm{p}\lambda}(\vec{r}) = e\vec{\alpha}\cdot\vec{A}_{\vec{q}_\mathrm{p}\lambda}(\vec{r})\,, \tag{9}$$

where $\vec{q}_\mathrm{p}$ and λ are the wave vector and polarization of the radiation with respect to the local frame of reference. This means that in general one has to transform $\vec{q}_\mathrm{p}$ and λ normally given in a global reference system to the local systems to calculate $M^{\vec{q}_\mathrm{p}\lambda}$.

VALENCE BAND-X-RAY PHOTO EMISSION

A relativistic theory for the spin- and angular resolved valence photo emission spectroscopy (VB-PES) on the basis of the one step model has been worked out by Feder and coworkers [12]. This very general scheme supplies a sound basis for an investigation of linear as well as the circular magnetic dichroism. Ebert and Schwitalla worked out a corresponding description for the angular integrated VB-PES and predicted the occurrence for circular dichroism also for this situation [13,14]. For the X-ray regime (XPS) multiple scattering events for the final states can be ignored leading to a rather simple expression for the spin (m_s) resolved photo emission intensity

$$I(E, m_s; \omega, \vec{q}_\mathrm{p}, \lambda) \propto \sum_q \Im \sum_{\substack{\Lambda\Lambda'' \\ \mu = \mu''}} C_\Lambda^{-m_s} C_{\Lambda''}^{-m_s}$$

$$\left\{\sum_{\Lambda_1\Lambda_2} \tau_{\Lambda_1\Lambda_2}^{qq}(E)\left[\sum_{\Lambda'} t_{\Lambda'\Lambda}^q(E')M_{\Lambda'\Lambda_1}^{\vec{q}_\mathrm{p}\lambda,q}\right] \times \left[\sum_{\Lambda'''} t_{\Lambda'''\Lambda''}^q(E')M_{\Lambda'''\Lambda_2}^{\vec{q}_\mathrm{p}\lambda,q}\right]^*\right.$$

$$\left. - \sum_{\Lambda'\Lambda'''\Lambda_1} t_{\Lambda'\Lambda}^q(E')I_{\Lambda'\Lambda_1\Lambda'''}^{\vec{q}_\mathrm{p}\lambda,q} t_{\Lambda'''\Lambda''}^{q*}(E')\right\}\,. \tag{10}$$

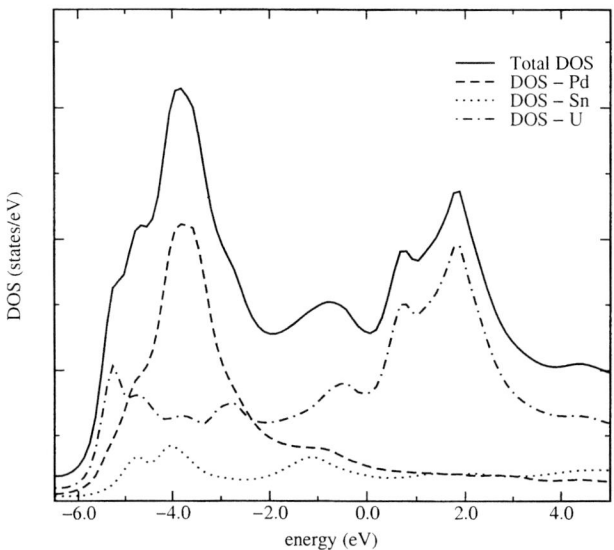

FIGURE 2. Component resolved density of states for UPdSn as calculated by the SPR-KKR method

For an ordered compound with many atoms per unit cell, one has to sum over the contributions stemming from the various atomic sites q within the unit cell.

Originally the above expression has been used with all quantities defined with respect to the global frame of reference. Application to the systems with a non-collinear spin structure, however, requires to evaluate all quantities with respect to the local frame of reference of an atomic site q. In particular one has to take into account that the polarization λ is specified in a global way. To describe experiments done with unpolarized radiation one has to average the intensity $I(E, m_s; \omega, \vec{q}_p, \lambda)$ with respect to the various polarization states λ. In this case it is not necessary to express the polarization states first with respect to the global system and then to transform to the local one because averaging with respect to the local polarization states leads to the same result.

The scheme sketched above has been applied to the compound UPdSn. Band structure calculations for this system have first been done by Trygg *et al.* [15] assuming a collinear anti-ferromagnetic structure. More recent work done by Sandratskii and Kübler [16] accounted for the non-collinear configuration and reproduced the experimental situation in a very satisfactory way. The electronic potential for UPdSn obtained in Ref. [16] has been used as an input to calculate the VB-XPS spectrum of UPdSn via the SPR-KKR-method on the basis of the finite cluster approximation. The corresponding density of states curves are given in Fig. 2. The partial density of states of U is dominated by its f-state contribution centered in

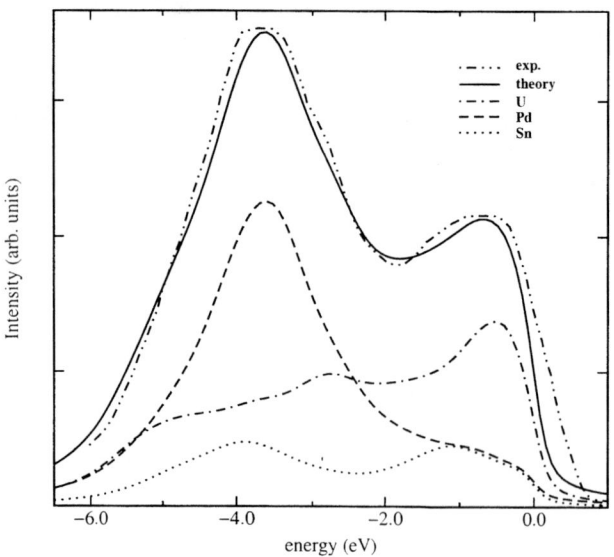

FIGURE 3. Component resolved VB-XPS-spectrum of UPdSn for the photon energy $E_{phot} = 1486.6$ eV. The total spectrum is compared with corresponding experimental data recorded by Havela et al. [17].

the vicinity of the Fermi level while the Pd density of states possesses a relatively narrow d-band complex at around 4 eV binding energy. The partial density of states of Sn is relatively featureless. These features are reflected in the partial contributions to the VB-XPS-intensity that are shown in Fig. 3. These curves have been calculated for a photon energy of 1486.6 eV and have been broadened to account for the various intrinsic and apparative broadening mechanisms. Here one notes that taking into account the partial cross sections or, equivalently, the matrix elements $M^{q_p, \vec{\lambda}, q}$ does not alter much the ratio between the component-resolved contributions compared with the partial density of states curves. To allow comparison with corresponding experimental data the partial spectra in Fig. 3 have been calculated assuming unpolarized radiation. The resulting total theoretical VB-XPS-spectrum in Fig. 3 is in rather good agreement with the spectra recorded by Havela et al. [17]. This clearly demonstrates that the approach outlined above is straightforwardly applicable to the magnetic systems possessing a non-collinear spin structure. In contrast to the ferromagnetic alloy systems Co_xPt_{1-x} [13] and Fe_xCo_{1-x} [14] studied before one cannot expect a circular dichroism to be found in the VB-XPS spectrum of UPdSn because of its spin compensated configuration (i.e. the net spin moment is zero). However, linear dichroism should be present and can be studied within the theoretical scheme discussed above.

X-RAY ABSORPTION

The magnetic dichroic effect observed in X-ray absorption can be described straightforwardly within the framework of the SPR-KKR formalism [18]. For the corresponding absorption coefficient $\mu^{\vec{q}_p \lambda q}(\omega)$ for a site q one may write:

$$\mu^{\vec{q}_p \lambda q}(\omega) \propto \sum_{i\,occ} \langle \Phi_i | X^{\times}_{\vec{q}_p \lambda} \, \Im G^+(E_i + \hbar\omega) \, X_{\vec{q}_p \lambda} | \Phi_i \rangle \, \theta(E_i + \hbar\omega - E_F) \,, \qquad (11)$$

where q_p, ω and λ stand for the wave vector, frequency and polarization of the radiation. The sum runs over all involved core states Φ_i. The itinerant final states above the Fermi level are again represented by the electronic Green's function within multiple scattering theory.

The above expression has been applied so far only to collinear spin systems. However, its extension to non-collinear systems is again very straightforward. Here we report the results of the calculations for the assumed spiral magnetic states in hcp-Gd. The spin-orbit coupling included into the calculational scheme destroys the generalized translational symmetry of the spiral structure and disturbs its perfect regularity. The effect of the influence of the spin-orbit coupling on the magnetic structure is not taken into account in the present model calculations.

In the calculations we used the self-consistent potentials obtained using the ASW-method. Here the calculation has been carried out for or a relatively small finite atomic cluster. (SPR-KKR calculations for ferromagnetic hcp-Gd using larger clusters of inappropriate size and comparison with experiment can be found in Ref. [19]).

The theoretical absorption spectrum for the L_2-edge of hcp-Gd obtained for various spin configurations are shown in Fig. 4. The top most panel shows for $\theta = 30°$ and various angles ϕ the spectra $\bar{\mu}_{L_2}$ for unpolarized radiation. Obviously there is hardly any influence of the spin structure on $\bar{\mu}_{L_2}$. As expected, this does not apply to the circular dichroic signal $\Delta\mu_{L_2} = \mu^+_{L_2} - \mu^-_{L_2}$ that is defined as the difference in absorption for left and right circularly polarized radiation. These spectra are given here with respect to the local frame of reference of an absorbing atom.

Fig. 4 clearly shows that $\Delta\mu_{L_2}$ is most affected by changes of the spin structure primarily in the near edge region. In particular a strong dependency on the angle θ can be seen. For $\theta = 30°$ there is hardly any dependency on ϕ found for the $\Delta\mu_{L_2}$ spectra. Increasing θ, changes of $\Delta\mu_{L_2}$ are more and more observed also with a variation of the angle ϕ.

Altogether, the changes on $\Delta\mu_{L_2}$ induced by varying the spin spiral structure by setting the angles θ and ϕ are – apart from the near edge region – astonishingly small. However, one has to keep in mind that the spectra refer to the local frame of reference. For the circularly polarized radiation coming along the crystallographic z-axis, the dichroic signal $\Delta\mu_{L_2}$ will be diminished accordingly by $cos(\theta)$.

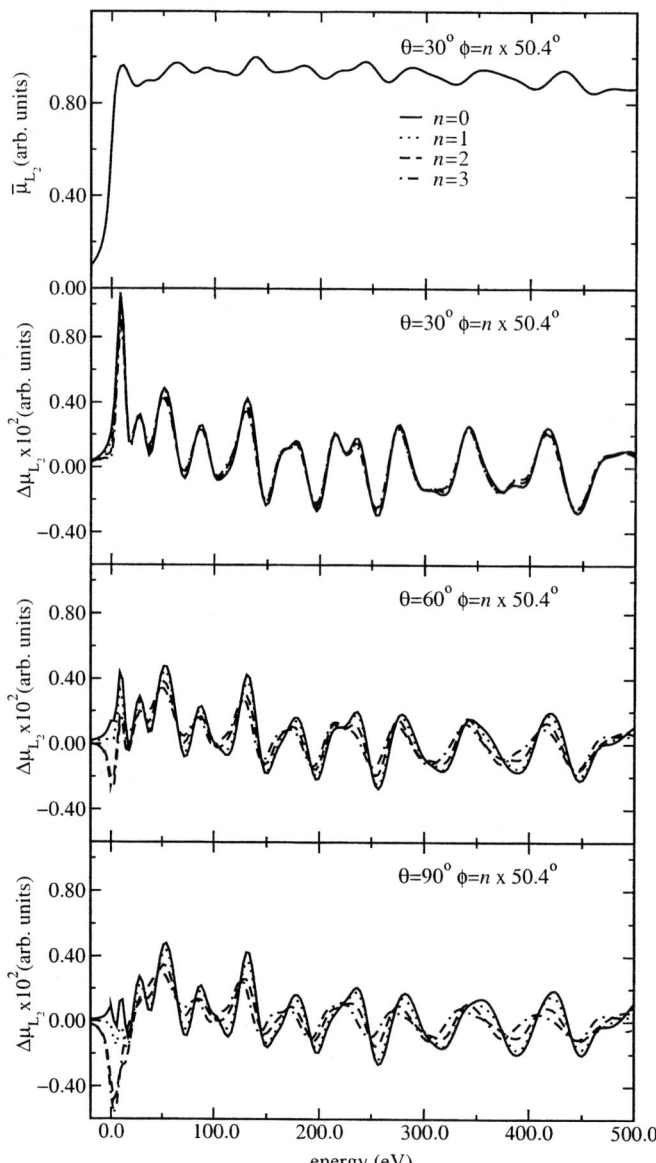

FIGURE 4. Theoretical X-ray absorption spectra for the L_2-edge of hcp-Gd. The top panel shows the polarization averaged absorption coefficient $\bar{\mu}_{L_2}$ while the other panels shows results for the magnetic circular dichroism defined as the difference in absorption for left and right circularly polarized radiation $\Delta\mu_{L_2} = \mu_{L_2}^+ - \mu_{L_2}^-$. The various spectra have been calculated assuming different spin spiral structures specified by the angles θ and ϕ (see text).

SUMMARY

The method to deal with non-collinear spin configurations within the framework of the relativistic SDFT has been briefly outlined. The single site problem is efficiently treated in the local frame of reference with the z-axis parallel to the spin magnetic moment. To solve the multiple scattering problem single site t-matrices are transformed from the local atomic systems to the global system. The method allows description of the variety of spectroscopic properties of non-collinear magnetic systems. The method is applied to the calculation of the VB-XPS spectrum of UPdSn and the magnetic dichroic spectra $\Delta\mu_{L_2}$ for assumed spiral magnetic states in hcp-Gd.

ACKNOWLEDGMENT

This work was funded by the BMBF (Bundesministerium für Bildung und Forschung) within the programme *Zirkular polarisierte Synchrotronstrahlung: Dichroismus, Magnetismus und Spinorientierung* under contract 05 SC8WMA 7.

REFERENCES

1. M. Freyss, D. Stoeffler, and H. Dreysse, Symposium. Mater. Res. Soc , 201 (1997).
2. L. M. Sandratskii, Adv. Phys. **47**, 91 (1998).
3. J. Kübler, K.-H. Höck, J. Sticht, and A. R. Williams, J. Appl. Physics **63**, 3482 (1988).
4. S. Mao, N. Amin, and E. Murdock, J. Appl. Physics **83**; 6807 (1998).
5. M. van Schilfgaarde, I. A. Abrikosov, and B. Johansson, **400**, 46 (1999).
6. M. E. Rose, *Relativistic Electron Theory*, Wiley, New York, 1961.
7. A. H. MacDonald and S. H. Vosko, J. Phys. C: Solid State Phys. **12**, 2977 (1979).
8. M. V. Ramana and A. K. Rajagopal, Adv. Chem. Phys. **54**, 231 (1983).
9. L. Nordström and D. J. Singh, Phys. Rev. Letters **76**, 4420 (1996).
10. H. Eschrig and V. D. P. Servedio, J. Comp. Chem. **20**, 23 (1999).
11. M. E. Rose, *Elementary Theory of Angular Momentum*, Wiley, New York, 1961.
12. S. V. Halilov, E. Tamura, D. Meinert, H. Gollisch, and R. Feder, J. Phys.: Condensed Matter **5**, 3859 (1993).
13. H. Ebert and J. Schwitalla, Phys. Rev. B **55**, 3100 (1997).
14. S. Ostanin and H. Ebert, Phys. Rev. B **58**, 11577 (1998).
15. J. Trygg, B. Johansson, and O. Eriksson, Phys. Rev. B **49**, 7165 (1994).
16. L. M. Sandratskii and J. Kübler, J. Phys.: Condensed Matter **9**, 4897 (1997).
17. L. Havela, T. Almeida, and J. R. Naegele, J. Alloys Comp. **181**, 205 (1992).
18. H. Ebert, Rep. Prog. Phys. **59**, 1665 (1996).
19. D. Ahlers, G. Schütz, V. Popescu, and H. Ebert, J. Appl. Physics **83**, 7082 (1998).

PART 2
PROBING THE STRUCTURAL PROPERTIES

A new recursive approach to photoelectron diffraction simulation

[1,2]F. J. García de Abajo, [1,3]M. A. Van Hove, and [1,3]C. S. Fadley

[1] Materials Sciences Division, Lawrence Berkeley National Laboratory, Berkeley, CA 94720, USA
[2] Departamento de CCIA (Facultad de Informática) and Donostia International Physics Center (DIPC), San Sebastián, Spain
[3] Department of Physics, University of California, Davis, CA 95616, USA

Abstract. A new recursive method for the simulation of photoelecton diffraction in solids within the cluster approach is presented. No approximations are made beyond the muffin-tin model, and in particular, an exact representation of the free-electron Green function is used. The new method relies upon a convenient separation of the free-electron Green function involving rotation matrices to reduce the computation time and storage demand. The multiple scattering expansion is iteratively evaluated using a divergence-free recursion method. The resulting computational demand scales as $N^2(l_{\max}+1)^3$ with the number of atoms in the cluster N and the maximum of the relevant angular momentum quantum numbers $l_{\max}$. Actual examples are given where $N > 1000$ is needed for convergence within 5% in the calculated photoelectron intensity.

INTRODUCTION

Multiple elastic scattering (MS) plays a central role in the description of electron transport inside solids in different experimental spectroscopies, and in particular in core-level photoelectron diffraction (PD) [1–3]. In this context, the relatively high electron energies usually employed (> 50 eV) permit approximating the atomic potentials by spherically-symmetric muffin-tin potentials [4]. Besides, inelastic scattering can be treated in a phenomenological way via complex optical potentials [4].

The cluster model adopted provides a natural approach to simulate MS effects in PD that is suggested by the fact that excited electrons cannot travel large distances in realistic solids without suffering inelastic losses, so that the region which actually contributes to the emission of elastically scattered electrons defines a finite cluster surrounding the emitter [2,5–9].

Typically, the electron wave function is expressed in spherical harmonics and spherical Bessel functions attached to each atom of the cluster in order to incorporate curved-wave effects. Unfortunately, the propagation of these functions between cluster atoms is computationally very demanding [10,11,4]. Therefore, approximations have been introduced in the past [12–14,5,15,16,7,8,17,18], some of them inspired in the high-energy limit, where the electron propagation reduces to plane-wave factors (plane-wave approximation) and each term in the MS series becomes a product of scattering amplitudes [19]. Expansions that take into consideration the finite size of the atoms have also been developed [5].

Beyond this, full curved-wave formulations of the problem require dealing with $(l_{max}+1)^2$ partial wave components per atom, where l_{max} is the maximum of the significant angular momentum quantum numbers, which scales roughly as $l_{max} \sim kr_{mt}$ with the electron momentum k and the atomic muffin-tin radius r_{mt}, and therefore, each propagation of the electron wave function between each pair of atoms involves $(l_{max}+1)^4$ complex products.

In order to overcome the rapidly-growing computational demand with increasing l_{max}, Rehr and Albers [7] (R-A) provided a clever procedure based upon a separable representation of the free-electron Green function that allows one to generalize the scattering amplitudes, substituting them by matrices that produce reliable results when keeping only a few of their leading elements [7]. This is particularly suitable to calculate the contribution of different individual electron paths.

In this paper, the MS expansion is evaluated using an exact representation of the Green function propagator. An iterative procedure is followed that requires $\approx (10/3)N^2(l_{max}+1)^3$ multiplications per iteration. Moreover, previously reported divergences in the MS series [20] are prevented by using Haydock's recursion method [21].

THEORY

Within the muffin-tin model adopted here, each atom in the cluster is represented by an atomic potential that vanishes outside a sphere of radius r_{mt}^α (the muffin-tin radius) centered at $\mathbf{R}_\alpha$. These are non-overlapping spheres and the total potential is set to a constant (the muffin-tin zero) in the interstitial region.

In core-level photoemission, the direct wave function ϕ^0 (i.e., before MS is considered) can be expressed as a combination of spherical outgoing waves centered around the emitter atom α_0:

$$\phi^0(\mathbf{r}) = \sum_L h_L^{(+)}[k(\mathbf{r}-\mathbf{R}_{\alpha_0})]\,\phi^0_{\alpha_0,L} \tag{1}$$

for $|\mathbf{r}-\mathbf{R}_{\alpha_0}| > r_{mt}^{\alpha_0}$, where $h_L^{(+)}(k\mathbf{r}) = i^l h_l^{(+)}(kr)Y_L(\Omega_\mathbf{r})$, $h_l^{(+)}$ is a spherical Hankel function [22], and $L = (l,m)$ labels spherical harmonics Y_L. The coefficients $\phi^0_{\alpha_0,L}$ depend on the geometry, polarization, and energy of the incident light and on the initial core state.

Single scattering of this direct wave leads to an extra contribution to the wave function than can also be expressed in terms of spherical outgoing waves centered around each of the cluster atoms. By iteratively employing this argument, one finds that the full self-consistent wave function ϕ has to be made of spherical waves as well (even after MS has been carried out up to an infinite order), so that

$$\phi(\mathbf{r}) = \sum_\alpha \sum_L h_L^{(+)}[k(\mathbf{r} - \mathbf{R}_\alpha)] \phi_{\alpha,L} \tag{2}$$

for $\mathbf{r}$ outside the muffin-tin spheres.

The coefficients $\phi_{\alpha,L}$ must be determined self-consistently by solving a secular equation that can be found as follows: for each atom α, $\phi_{\alpha,L}$ is the sum of the direct wave (only contributing for α_0, that is, the emitter) plus the result of the free propagation of the electron curved-wave components from every other atom β up to atom α, followed by scattering at the latter. More precisely,

$$\phi_\alpha = \phi_\alpha^0 + t_\alpha \sum_{\beta \neq \alpha} G_{\alpha\beta}\phi_\beta, \tag{3}$$

where $G_{\alpha\beta}$ takes care of the abovementioned propagation, t_α is the scattering matrix of atom α, $\phi_\alpha^0 = \delta_{\alpha\alpha_0}\phi_{\alpha_0}^0$, and ϕ_α is the vector of components $\phi_{\alpha,L}$. In the basis set of spherical harmonics attached to each cluster atom, one has $t_{\alpha,LL'} = \sin \delta_l^\alpha e^{i\delta_l^\alpha} \delta_{LL'}$, where δ_l^α is the lth scattering phase shift of atom α [22]. Besides, the matrix $G_{\alpha\beta}$ is connected to the translation formula of spherical harmonics and its detailed expression can be written

$$G_{\alpha\beta,LL'} = 4\pi \sum_{L''} h_{L''}^{(+)}(k\mathbf{d}_{\alpha\beta}) \int d\Omega\, Y_L(\Omega) Y_{L''}(\Omega) Y_{L'}^*(\Omega), \tag{4}$$

where $\mathbf{d}_{\alpha\beta} = \mathbf{R}_\alpha - \mathbf{R}_\beta$, for $|\mathbf{r} - \mathbf{R}_\alpha| < d_{\alpha\beta}$ (this condition is satisfied when $\mathbf{r}$ is contained inside the muffin-tin sphere $\beta \neq \alpha$ and non-overlapping spheres are considered). Inelastic effects are described via either an imaginary part in k or, to an excellent approximation, an exponential factor $\exp(-d_{\alpha\beta}/2\lambda_i)$.

The photoelectron intensity at the detector can be directly obtained from Eq. (2) in the $kr \to \infty$ limit. One finds

$$I(\hat{\mathbf{r}}) \propto |\sum_\alpha e^{-i\mathbf{k}_f \cdot \mathbf{R}_\alpha - \zeta_\alpha/2\lambda_i} \sum_L Y_L(\Omega)\phi_{\alpha,L}|^2, \tag{5}$$

where $\mathbf{k}_f = k\mathbf{r}/r$ and ζ_α, the distance from atom α to the surface along the direction of emission, is introduced to account for inelastic attenuation of the photoelectron before it leaves the solid.

Three different iterative techniques have been used and compared in the present work to evaluate Eq. (3): (a) direct Jacobi iteration, consisting in starting with $\phi \approx \phi^0$ and using Eq. (3) as an iteration formula to improve ϕ (this results in the intuitive MS procedure where every iteration leads to the next order of scattering); (b) simultaneous relaxation (SR) [23], previously used in this context [20], and consisting in both using the latest values of the coefficients of ϕ as soon as they are calculated and mixing the result of each iteration with the previous one to improve convergence (the fractional weight of the former will be denoted η); and (c) Haydock's recursion method [21], modified in a way suitable to obtain photoemission intensities along an arbitrary number of directions of emission with a single MS calculation for each electron energy, as will be discussed elsewhere [24].

Rather than directly using Eq. (4), $G_{\alpha\beta}$ will be constructed in three steps as follows [10,7]: first, the bond vector $\mathbf{d}_{\alpha\beta}$ is rotated onto the z axis by using a rotation matrix $R_{\alpha\beta}$ [22,7]; the resulting rotated wave function components are then propagated a distance $d_{\alpha\beta}$ along the positive direction of the z axis by multiplying by $G^z_{\alpha\beta}$, calculated from Eq. (5) by using $(0, 0, d_{\alpha\beta})$ instead of $\mathbf{d}_{\alpha\beta}$; finally, the z axis is rotated back onto the $\mathbf{d}_{\alpha\beta}$ direction, and one has

$$G_{\alpha\beta} = R_{\alpha\beta}^{-1} G^z_{\alpha\beta} R_{\alpha\beta}. \tag{6}$$

The rotation matrices involved here can be in turn decomposed into azimuthal and polar rotations.

A significant reduction in memory demand can be accomplished if the coefficients of each polar rotation, each azimuthal rotation, and each propagation $G^z_{\alpha\beta,LL'}$ are computed and stored once and for all the first time that they are encountered during the full calculation. Besides, all of the matrices that appear on the right hand side of Eq. (6) are sparse, and a detailed inspection leads to the conclusion that the number of complex multiplications needed to evaluate each product $G_{\alpha\beta}\phi_\beta$ when using this decomposition is cut down by a factor of $\approx 3l_{\max}/10$.

EXAMPLES AND DISCUSSION

The performance of the various iteration methods discussed above to calculate PD from a simple sample consisting of two carbon atoms is compared in Fig. 1, where the inset illustrates the details of the geometry (the interatomic distance corresponds to nearest neighbors in graphite). Scattering from a cluster of carbon atoms is a severe test case, as multiple scattering between bonded carbon atoms is particularly strong. Within the resolution of the figure, the recursion method (solid circles) converges in just seven iterations. At single scattering, the direct Jacobi iteration (open circles) lies 4% off the exact result, and subsequent scattering orders lead to divergence. The latter is not prevented by using the SR method (broken curves) over a wide range of the relaxation parameter η. The lower η, the slower the increase in intensity with iteration step, but the divergent behavior remains.

As is well known in LEED [4], divergences like this one are encountered in MS when the absolute value of any of the eigenvalues of the eigensystem (3) is larger than 1. The SR method provides a cure in many cases [20], but it is not sufficiently general, as illustrated by Fig. 1. Instead, the recursion method has a well-established convergent behavior [21].

The efficiency of our new method allows us to perform calculations for much larger clusters than with other methods (e.g., we have calculated full-hemisphere distributions for a cluster of about 2500 atoms [24]). We can thus investigate the question of cluster size convergence, as is done in Fig. 2 for photoemission from a Cu2s level situated on the third layer of a Cu(111) surface. The geometry under consideration is illustrated schematically on the lower left corner of the figure. Plotted here is the reliability factor defined as

$$R = \frac{\overline{|I^N - I^\infty|}}{\overline{I^\infty}}, \tag{7}$$

where the average is taken over azimuthal directions of emission, I^N is the intensity calculated for an N−atom cluster, and I^∞ is actually obtained for $N = 1856$. The solid curve and circles correspond to the result obtained from the recursion method, where convergence is achieved in less than 20 iterations. A smooth convergence can be seen in the $N \to \infty$ limit. The inset shows azimuthal scans obtained for different cluster sizes, in order to facilitate the understanding of the actual meaning of R in terms of curve comparisons. For $N = 944$ (dotted curve in the inset), one has

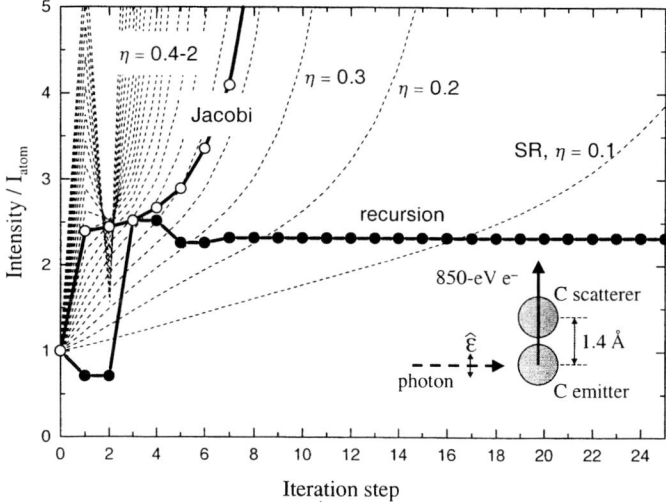

FIGURE 1. C1s photoemission intensity in a cluster formed by two carbon atoms separated by 1.4 Å as a function of the number of iteration steps. The incoming light is linearly polarized with the polarization vector parallel to the interatomic axis. The emission occurs in the forward-scattering direction (see inset). The electron energy is 850 eV. Results obtained from different iteration methods are compared: the recursion method (solid curve); the direct Jacobi iteration (dotted and dashed curves); for which the number of iteration steps equals the scattering order; and the simultaneous relaxation for various values of the relaxation parameter η (thin broken curves). The intensity has been normalized to that of the isolated C atom.

$R = 0.03$ and convergence is already quite good as compared to the $N = 1856$ case, although some small discrepancies can still be distinguished in the height of the peaks around 30°, 60°, and 90°, so that over 1000 atoms are needed to obtain convergence within the resolution of the figure (deviations above 5% in intensity are observed for some directions of emission using $N = 944$).

The open circles in Fig. 2 show the reliability factor obtained from the Jacobi method for various scattering orders (5, 9, 13, 17, 21, and 25), where the spread in the position of the circles makes evident a divergent behavior. The latter is more pronounced for larger clusters. In this sense, the Jacobi method has to be regarded as an asymptotic series unable to converge below a certain reliability factor in the present case.

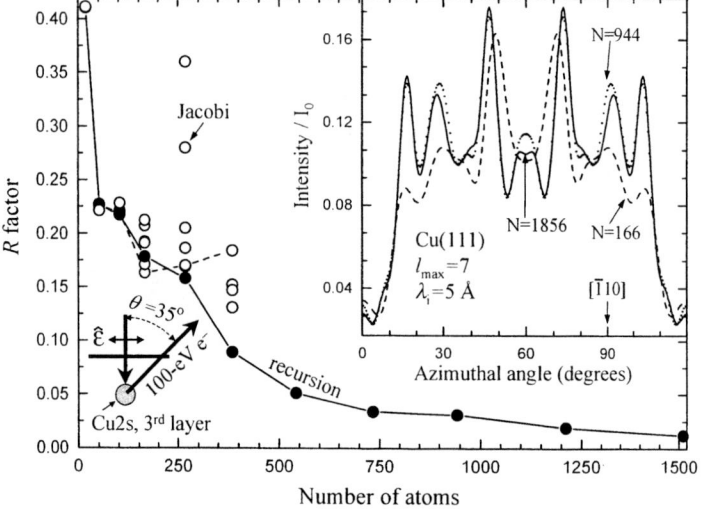

FIGURE 2. R–factor [Eq. (7)] variation with the number of atoms N for Cu2s photoemission from the third layer of a Cu(111) surface. Azimuthal scans have been considered with a polar angle of emission of 35°, a photoelectron energy of 100 eV, and p-polarized light under normal incidence conditions, as shown schematically on the lower left corner of the figure. The inset shows the intensity as a function of azimuthal angle for various cluster sizes, as indicated by labels, normalized to that of the direct emission without inelastic attenuation.

ACKNOWLEDGMENTS

This work was supported in part by the University of the Basque Country and the Spanish Ministerio de Educación y Cultura (Fulbright grant FU-98-22726216), and in part by the Director, Office of Science, Basic Energy Sciences, Materials Sciences Division, of the U.S. Department of Energy under Contract No. DE-AC03-76SF00098.

REFERENCES

1. C. S. Fadley and S. Å. L. Bergström, Phys. Lett. A **35**, 375 (1971).
2. A. Liebsch, Phys. Rev. Lett. **32**, 1203 (1974).
3. D. P. Woodruff et al., Phys. Rev. Lett. **41**, 1130 (1978).
4. J. B. Pendry, *Low Energy Electron Diffraction* (Academic Press, London, 1974).
5. J. J. Barton and D. A. Shirley, Phys. Rev. A **32**, 1019 (1985).
6. J. J. Barton, M.-L. Xu, and M. A. Van Hove, Phys. Rev. B **37**, 10475 (1988).
7. J. J. Rehr and R. C. Albers, Phys. Rev. B **41**, 8139 (1990).
8. V. Fritzsche, J. Phys.: Condens. Matter **2**, 9735 (1990).
9. D. J. Friedman and C. S. Fadley, J. Electron Spectrosc. **51**, 689 (1990).
10. M. Danos and L. C. Maximon, J. Math. Phys. **6**, 766 (1965).
11. R. Nozawa, Nucl. Instrum. Methods B **100**, 1 (1966).
12. W. L. Schaich, Phys. Rev. B **8**, 4028 (1973).
13. C. A. Ashley and S. Doniach, Phys. Rev. B **11**, 1279 (1975).
14. W. L. Schaich, Phys. Rev. B **29**, 6513 (1984).
15. S. J. Gurman, N. Binsted, and I. Ross, J. Phys.: Condens. Matter **19**, 1845 (1986).
16. J. Mustre de Leon et al., Phys. Rev. B **39**, 5632 (1989).
17. A. P. Kaduwela, D. J. Friedman, and C. S. Fadley, J. Electron Spectrosc. **57**, 223 (1991).
18. Y. Chen et al., Phys. Rev. B **58**, 13121 (1998).
19. P. A. Lee and J. B. Pendry, Phys. Rev. B **11**, 2795 (1975).
20. H. Wu and S. Y. Tong, Phys. Rev. B **59**, 1657 (1999).
21. R. Haydock, Solid State Physics **35**, 215 (1980).
22. A. Messiah, *Quantum Mechanics* (North-Holland, New York, 1966).
23. W. H. Press, S. A. Teukolsky, W. T. Vetterling, and B. P. Flannery, *Numerical Recipes* (Cambridge University Press, New York, 1992).
24. F. J. García de Abajo, M. A. Van Hove, and C. S. Fadley, (in preparation).

Direct Methods for Surface X-Ray Diffraction

D. K. Saldin*, R. Harder,* V. L. Shneerson*, H. Vogler^S and W. Moritz^S

*Department of Physics and Laboratory for Surface Studies
University of Wisconsin-Milwaukee
P. O. Box 413, Milwaukee, WI 53201, U.S.A.
^{S}Institute for Crystallography and Mineralogy
University of Munich
Theresienstrsse 41, 80333 Munich, Germany

Abstract. We develop of a direct method for surface X-ray diffraction that exploits the holographic feature of a known reference wave from the substrate. A Bayesian analysis of the optimal inference to be made from an incomplete data set suggests a maximum entropy algorithm that balances agreement with the data and other statistical considerations.

INTRODUCTION

The basic aim of X-ray crystallography is reconstruct the electron density distribution of a unit cell from a knowledge of measured diffracted intensities. Although the diffracted *amplitudes* are proportional to the square roots of the measured *intensities*, and thus known directly from experiment, there is no easy way to measure their *phases*. If the latter quantities were also known, recovering the electron density would be a simple matter of Fourier transformation, which can be performed easily on a computer. It is the lack of direct knowledge of these phases that is the well-known *phase problem* of crystallography.

Consequently, one method of structure solution that has been employed extensively in surface crystallography is to work in the reverse direction: the diffracted intensities resulting from a model of the structure are calculated and compared with the measured intensities. The degree of agreement is quantified by a quantity called a reliability factor (or R-factor), and the process repeated with other models of the structure in some systematic fashion. The model that yields the lowest value of the R-factor is assumed to be the correct solution.

A couple of limitations of this approach are: (a) the models are chosen subjectively, and may be limited by human imagination; (b) the capabilities of present or even any conceivable future computer may be insufficient to consider even all possible imagined structure models in a reasonable time [1]. Consequently, there is an interest in developing a so-called *direct* algorithm in which the electron density is recovered directly from the diffraction data from objective considerations, with essentially no human intervention. In this paper we compare the application of two classes of direct methods for the problem of surface X-ray diffraction. One seeks to obtain the electron density distribution that yields the best fit to a given set of experimental data. We will show that, in this approach, the limitations of the experimental data can sometimes give rise to artificial and misleading electron density distributions. The second approach, which takes advantage of the principles of Bayesian statistics [2], allows a better reconstruction of the electron density by sacrificing some degree of fit to the experimental data in favor of other prior information. We begin with a description of the experiment and the problem to be solved.

SURFACE X-RAY DIFFRACTION

The technique of surface X-ray diffraction attempts to recover the electron density of a surface from an analysis of the measured scattered amplitudes from a collimated beam of X-rays of glancing incidence. A reciprocal space diagram of the experimental geometry is shown in Fig. 1. The wavevector of X-rays of glancing incidence is denoted by k_0, and the wavevector of the detected beam is k. Due to the 2D periodicity of the surface, the wavevector difference, $q=k-k_0$ must join the reciprocal space origin (at the end of the wavevecor k_0) with the intersection of a rod specified by in-plane Miller indices (H,K) and the Ewald sphere (at out-of-plane Miller index value L). Thus the vector q may be specified by the three Miller indices (H,K,L). Although the 2D periodicity restricts the values of H and K to be integers, the lack of periodicity perpendicular to the surface allows non-zero diffracted intensity for arbitrary values of L, which may be varied by rotating the sample relative to the incident beam as shown. The intensity variation with L of the diffracted amplitudes corresponding to particular integer values of H and K is known as a crystal truncation rod (CTR).

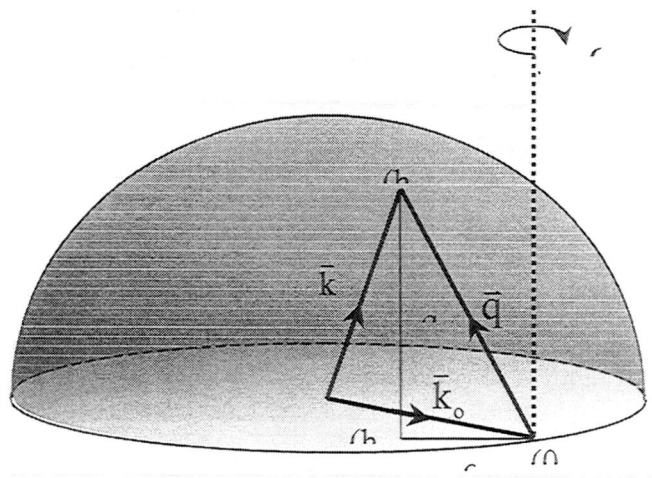

FIGURE 1.

The intensity of a beam corresponding to the scattered wavevector q is essentially the square modulus of the structure factor F_q^{exp}. Thus the *modulus* of the structure factor may be determined immediately from the measured data, but not its phase. If

(1)

the latter were also known, it would be a simple matter to recover the surface electron density u_i at position r_i by the Fourier transform

$$u_i = \frac{1}{N}\sum_q \left(F_q^{exp} - B_q\right)\exp(-iq.r_i)$$

where B_q is the (known) diffracted amplitude from the bulk, and N the number of values of q included in the sum in (1). Given the lack of this phase information, a more sophisticated algorithm is required to reconstruct the electron density.

HOLOGRAPHIC METHOD

One way to recover the surface electron density is by analogy with holography [3]. The scattered amplitude (structure factor) may be written in the form:

$$F_q^{calc} = B_q + S_q \tag{2}$$

a sum of the known contribution, B_q, from the bulk (which may be identified with a reference wave) and an unknown surface contribution (representing an object wave).

$$S_q = \sum_j u_j \exp(iq.r_i) \tag{3}$$

Substituting (3) into (2), and equating the measured intensity $|F_q^{exp}|^2$ to its calculated counterpart yields

$$\left|F_q^{exp}\right|^2 - \left|B_q\right|^2 = \sum_j u_j \left\{ M_{q,j} + \sum_l u_l M_{q,j}^* M_{q,l} \right\} \tag{4}$$

where

$$M_{q,j} = \exp(iq.r_j) \tag{5}$$

It will be noted that the LHS of Eq.(4) represents a difference between a measurable quantity $|F_q^{exp}|^2$ and a known one $|B_q|^2$, while all quantities on the RHS are known, except for the surface electron distribution $\{u_j\}$. If the quantities $|F_q^{exp}|$ are measured for a sufficient number of different scattering vectors q, the resulting set of equations (4) may be solved to recover that electron distribution [4].

We illustrate this with a test calculation to recover the (laterally-averaged) variation, in a direction perpendicular to the surface of Ag(001), of the charge density of a K atom adsorbate at a height of 4.29 Å above the outermost substrate layer. This requires only a measurement of the (00L) (specular) CTR, and a set of bulk diffracted amplitudes B_q, which may be calculated, since the bulk structure is assumed known. For our test we simulated the "experimental" data, $|F_{00L}^{exp}|$, shown in Fig. 2, using the ROD computer program of Vlieg [5]. We then recovered the electron density $\{u_j\}$ that minimizes the disagreement between the LHS and RHS of Eq. (4) by means of the APRIORI scheme of Saldin et al. [6], which employs repeated applications of a linear programming algorithm [7], to solve the non-linear equations (4). The progress of the recovery may be monitored by calculating the X-ray R-factor, R1 [8], quantifying the degree of agreement between the "experimental" data and a simulation of it from the recovered electron distribution. The iterations were halted when R1 reached about

0.06, an excellent R-factor by any standards. The corresponding electron distribution is shown in Fig. 3.

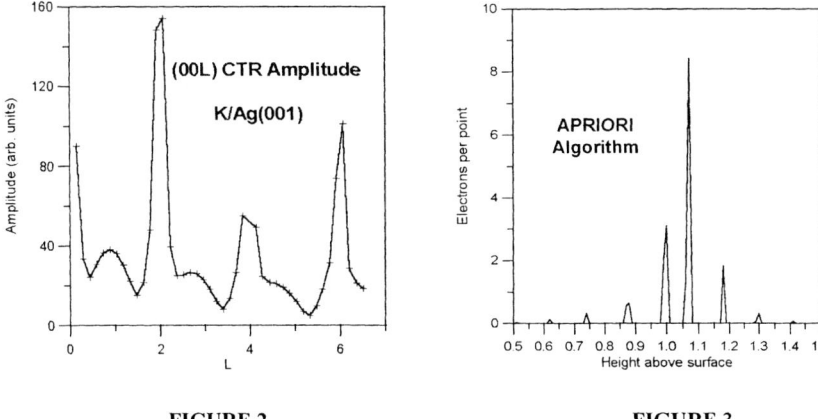

FIGURE 2. **FIGURE 3.**

The abscissa in Fig. 3 represents distance above the outermost substrate layer in units of the bulk repeat length, 4.086 Å, in this direction. In these units the height of the K atom assumed in our model is 1.05. The resulting electron distribution has a spiky unphysical appearance, although there does appear to be a cluster of strong spikes in the vicinity of the adsorbate. The mean spacing of the main spikes in Fig. 3 is what would be expected from such a finite range of diffraction data, which limits the resolution of the reconstructed real-space distribution. The smoothness of the real electron density distribution is not reconstructed by such an algorithm despite the near-perfect R-factor of the final model.

It is clear that in the presence of limited (and possibly noisy) data, it is counter productive to over-fit a model to the data. As we will see, an approach based on Bayesian methods of statistical inference is more appropriate in such a case. We begin with a discussion of that theory.

X-RAY CRYSTALLOGRAPHY AND BAYES' THEOREM

A theorem due to the 18th century mathematician Thomas Bayes [2] is highly relevant to the problem of inferring an electron density distribution from a set of scattered X-ray amplitudes. For our present problem of recovering an electron density distribution, ρ, from a set of measured diffraction "data", it may be stated:

$$\Pr ob(\rho \mid "data", I) = \frac{\Pr ob("data" \mid \rho, I) \times \Pr ob(\rho \mid I)}{\Pr ob("data" \mid I)} \qquad (6)$$

The LHS of the above equation may be regarded as the *posterior probability* of the electron density given the experimental data and background information. The first term in the numerator on the RHS is known as the *likelihood function* of the data for a given model of the electron density and the background information, and the second term in the numerator the *prior probability* of the electron density given the background information only. The denominator on the RHS is known as the *evidence*. For our problem the last-named quantity may be taken as constant, implying that all measured data is to be treated on an equal footing, and not biased by any background information. Eq.(4) then reduces to:

$$\Pr ob(\rho \,|\, "data", I) \propto \Pr ob("data"\,|\, \rho, I) \times \Pr ob(\rho \,|\, I) \quad (7)$$

The probability on the LHS of the proportionality (7) is what is sought in X-ray crystallography, and the above expression indicates that it is proportional to product of the likelihood function and the prior probability.

In terms of a climatic analogy, if we make the identifications:

$$"data" \Leftrightarrow \text{overhead clouds}$$

$$\rho \Leftrightarrow \text{rain}$$

Eq.(6) tells us that the probability of overhead clouds when its raining is not equal to the probability of rain when there are clouds overhead. From experience we would judge the former probability to be close to unity, but the latter could be quite small. However, the two quantities are related by the prior probability on the RHS.

In the context of X-ray crystallography, Prob("data"|ρ,I) is the probability that a particular electron density model produces the measured intensities, a quantity that is usually estimated by conventional trial-and-error methods. On the other hand the quantity Prob(ρ|"data",I), the probability of a particular electron density for a given data set, is the quantity actually sought in crystallography. Eq.(7) shows the inequality of these quantities, but it also shows how to calculate one from the other when armed with a knowledge of the prior probability.

In the next two sub-sections we show how to calculate the likelihood function and the prior probability in the case of X-ray crystallography. The quantity of interest, namely the most probable electron density consistent with the measured diffraction data and any background information, may then be found by maximizing the functional formed by the product of the two probabilities on the RHS of (7).

Likelihood function

The likelihood function may be determined by the following argument: given an estimate σ_g of the standard deviation of the error in the experimental measurement of a structure factor amplitude F_q^{exp}, the probability of such a measurement for a given model of the structure with calculated amplitude F_q^{calc} is given by the Gaussian distribution:

$$\Pr ob(F_q^{exp} \,|\, \rho, I) = \frac{1}{\sigma_q \sqrt{2\pi}} \exp\left\{ -\frac{|F_q^{exp} - F_q^{calc}|^2}{\sigma_q^2} \right\} \quad (8)$$

Given the multiplicative nature of independent probabilities, it follows that the likelihood function for the entire set of measured data is given by:

$$\Pr ob("data"\,|\, \rho, I) = \Pr ob(\{F_q^{exp}\} \,|\, \rho, I) \propto \exp\left\{ -\frac{\chi^2}{2} \right\} \quad (9)$$

where

$$\chi^2 = \frac{1}{N} \sum_q \frac{|F_q^{\exp} - F_q^{calc}|^2}{\sigma_q^2} \quad (10)$$

This "chi-squared statistic" [2] could be regarded as another form of R-factor. The quantity F_q^{calc} is essentially the square root of the measured intensity. From the definition of the standard deviation σ_q, the minimum expected value is of χ^2 would be of the order of unity.

Eq.(9) shows that minimization of this R-factor is equivalent to maximizing the likelihood function. Eq.(10) shows, although this may be the standard practice in conventional trial-and-error structure determinations, the R-factor is only one element in the correct determination of a structure from a given set of measured data. It is necessary also to take account of the prior probability, as we will now show.

Prior Probability

The prior probability of the electron density is simply the probability of a distribution given the background information, which is, in this case, just the positivity of the electron density and any prior estimate of that density on e.g. chemical grounds. Any known periodicity allows attention to be confined to the electron density within a repeat unit (or unit cell). The *a priori* probability of such a distribution is analogous to the probability P(u|I) that a team of hypothetical monkeys (our archetypal unbiased individuals, see Fig. 4) throw objects at random into a set of boxes 1, 2, 3, ... of capacities m_1, m_2, m_3, ... to produce a distribution, u_1, u_2, u_3, ..., of those objects (where *I* now includes the information about the capacities of the boxes).

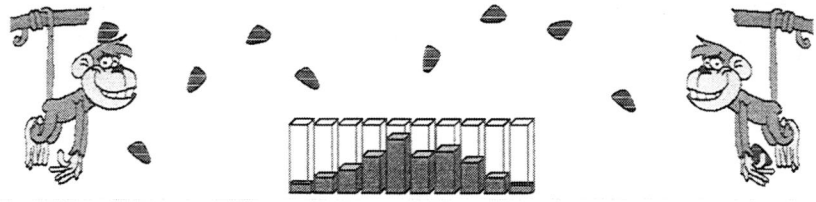

FIGURE 4.

From the standard combinatorial argument,

$$\operatorname{Pr}ob(u \mid I) \propto \Omega = \prod_i m_i^{u_i} \Big/ \prod_i u_i! \quad (11)$$

where Ω is the number of ways the distribution may come about. From Boltzm relation, the entropy, S, of the distribution may be written

$$S = \alpha \ln(\Omega) \quad (12)$$

where α is a constant. Combining Eqs. (11) and (12), we may write the prior probability as

$$\operatorname{Pr}ob(u \mid I) \propto \exp(S/\alpha) \quad (13)$$

In the crystallographic application, the quantities u_i are identified with the number of electrons in voxel (or "volume pixel") associated with grid point i within a unit cell. Thus the distribution $\{u\}$ is proportional to the electron density $\{\rho\}$ that crystallographers seek to determine. The distribution $\{m\}$ may be regarded as a prior estimate of $\{u\}$ based on the background information alone. For example, in protein crystallography, if the general shape of the molecule is known to low resolution, e.g. through a previously determined *solvent mask*, that knowledge may be incorporated into $\{m\}$.

Posterior Probability

Combining Eqs.(7), (9), and (13), we find that the posterior probability may be written as

$$\text{Pr}ob(u\,|\,"data",I) \propto \exp\left(\frac{S}{\alpha} - \frac{1}{2}\chi^2\right) \tag{14}$$

MAXIMUM ENTROPY METHOD

The most likely electron density distribution consistent with the experimental data and the background information is thus that which maximizes the exponent on the RHS of Eq.(17), or equivalently (since α is arbitrary), the functional, $Q[\{u_i\}]$, where:

$$Q[\{u_i\}] = \frac{S[\{u_i\}]}{\alpha} - \frac{\lambda}{2}\chi^2[\{u_i\}] \tag{15}$$

Regarding λ as a Lagrange multiplier, we see that the problem of finding the most likely electron density consistent with the data may be reduced to that of finding the distribution of maximum entropy constrained by the requirement of minimizing the "chi-squared" statistic. This is exactly the procedure of the maximum entropy method (MEM) of Jaynes [9].

The MEM has been applied to the problem of improving the initial "experimental" estimate of a set of phases of Bragg reflections in macromolecular crystallography by Collins [10]. There have been several subsequent applications of the MEM in this field (for a recent review, see Gilmore [11]). We present here an application of this method to surface X-ray diffraction (SXRD).

From Eqs. (11) and (12), and using Stirling's approximation:

$$\ln(u!) = u\ln(u) - u \tag{16}$$

it follows that the entropy term in Eq.(15) may be written as

$$\frac{S}{\alpha} = -\sum_i u_i \ln(u_i/em_i) \tag{17}$$

Substituting Eqs. (17) and (10) into Eq. (15), and taking $\sigma_q=1$ for all **q** (for theoretical "data") yields

$$Q[\{u_i\}] = -\sum_i u_i \ln\left(\frac{u_i}{em_i}\right) - \frac{\lambda}{2}\sum_q \left||F_q^{exp}|\exp(i\phi_q) - B_q - \sum_i u_i \exp(iq.r_i)\right|^2 \quad (18)$$

According to the theory developed above, the most probable electron density distribution consistent with the experimental data $|F_q^{exp}|$ is that which maximizes the functional Q. This may be found by setting

$$\frac{\partial}{\partial u_i} Q[\{u_i\}] = -\ln\left(\frac{u_i}{m_i}\right) - \lambda(t_i - u_i) = 0 \quad (19)$$

for all i. This leads to the equation

$$u_i = m_i \exp\{-\lambda(u_i - t_i)\} \quad (20)$$

where

$$t_i = \frac{1}{N}\sum_q \{|F_q^{exp}|\exp(i\phi_q) - B_q\}\exp(iq.r_i) \quad (21)$$

Since our purpose is to find the unknown surface electron density $\{u_i\}$ we need to find a solution to Eq.(20). Since the phases $\{\phi_q\}$ are also unknown, we need to find an iterative self-consistent algorithm. A similar algorithm was proposed by Collins [9] for the slightly different problem of improving the "experimental" phases in protein structure determination. It has been extended to the problem of protein structure completion by Saldin et al. [12], which is formally equivalent to the surface structure determination problem discussed here. The reason is that the diffracted amplitudes, e.g. F_q^{exp}, in surface crystallography are always a sum of a contribution B_q from a known bulk, and an unknown contribution, to be determined, from the surface.

Practical Algorithm

It will be noted that the "target function" t_i has the form expected of the surface electron density: its consists of the difference between the total diffracted amplitude

$$u_i^{(n)} = u_i^{(n-1)} \exp\{-\lambda(u_i^{(n-1)} - t_i^{(n-1)})\} \quad (22)$$

from the surface and that from the bulk. The only unknowns are the phases $\{\phi_q\}$ of the measured data. If the RHS of Eq. (20) may be thought of as an expression for improving a previous estimate of the quantity u_i, its self-correcting nature *vis a vis* the target function becomes clear: if the previous estimate of u_i is greater than t_i, Eq.(20) would tend to reduce u_i relative to its original estimate, m_i. This suggests that in an iterative algorithm, m_i should be identified with the estimate of u_i at the previous iteration. Conversely, if u_i were less than t_i on the RHS, it would tend to increase the earlier estimate of u_i. The combination of these arguments suggest the recasting of Eq.(20) into the iterative form:
An initial estimate of the surface electron density $\{u_i^{(0)}\}$ at the first iteration, n=1, may be just a uniform distribution normalized to the expected total number of electrons in the surface unit cell. As for $t_i^{(0)}$, that requires an initial estimate of the phase in

Eq.(21), $\phi_q^{(0)}$, say. We have found that an adequate estimate of this initial phase is the argument of the amplitude from a surface represented by a truncated bulk structure. This is easily calculated since the bulk structure is assumed known. At every subsequent iteration, this phase is calculated from:

$$\phi_q^{(n-1)} = \arg\left(B_q + \sum_i u_i^{(n-1)} \exp(iq.r_i)\right)$$

Note that the process is begun with an initial estimate of the phases of the measured diffraction amplitudes in reciprocal space. A Fourier transform yields the initial real-space target function $\{t_i\}$, which, in turn, is used to produce the next estimate of the electron distribution $\{u_i\}$. An inverse Fourier transform then returns an improved estimate of the phases $\{\phi_q\}$. In terms of its repeated oscillation between real and reciprocal space this procedure has something in common with the Gerchberg-Saxton algorithm [13], which is also used for recovering phases from a real-space amplitude distribution. In reciprocal space the total diffracted amplitudes are kept at their experimentally determined values. In real space the electron distribution may be confined to its estimated physical bounds. The imposition of these restrictions at each iteration is one feature that enables the algorithm to successively improve its estimate of the unknown crystallographic phases.

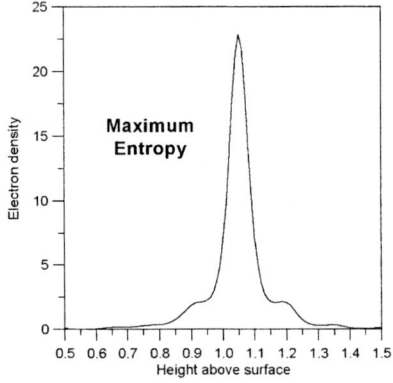

FIGURE 5.

Test Case

The 1D electron density distribution reconstructed by this maximum entropy algorithm from the CTR data in Fig.2 (for the K/Ag(001) surface) is shown in Fig. 5. Note the much smoother distribution expected of a real electron distribution, peaked exactly at the atom height above the surface. Yet the X-ray R-factor, R1, quantifying the agreement between the CTR expected from this distribution and the "experimental" CTR is only 0.28, signifying a significantly worse agreement than that from the unphysical spiky distribution of Fig. 3, recovered by the APRIORI scheme.

DISCUSSION AND CONCLUSIONS

With the advent of a new generation of synchrotron radiation sources, surface X-ray diffraction is emerging as one of the most attractive methods for surface crystallography. The relatively weak scattering of X-rays from a surface allows the diffracted amplitude to be written as a Fourier transform of the surface electron density. In turn, this allows the development of a direct method that exploits the fact that a major part of the scattered amplitude is one that comes from the known bulk structure, a holographic feature that allows the phases of the scattered waves to be determined by an appropriate numerical algorithm.

We have found that an algorithm whose sole charge is to find an electron distribution that best fits the data can give rise unphysical distributions. Instead, one that balances the need to fit incomplete data with other requirements based on prior statistical knowledge has been shown to be more successful in recovering the true electron distribution of the surface, despite a higher (or worse) R-factor. While the illustrative example in this paper is the relatively simple one of recovering a 1D electron density distribution from a specular crystal truncation rod, we have also confirmed that the method is remarkably successful in reconstructing fully 3D electron distributions of surfaces that may include relaxed and reconstructed substrates. The latter work will be reported in future publications.

REFERENCES

1. Pendry, J. B., Heinz, K., and Oed, W., *Phys. Rev. Letters* **61**, 2953-2956 (1988).
2. Sivia, D. S., *Data Analysis: A Bayesian Tutorial*, Oxford: Oxford University Press, 1996.
3. Collier, R. J., Burckhardt, C. B., and Lin, L. H., *Optical Holography*, New York: Academic, 1971.
4. Szöke, A. *Phys. Rev. B* **47**, 14 044-14 048 (1993).
5. E. Vlieg, J. Appl. Cryst, in press.
6. Saldin, D. K., Chen, X., Kothari, N., and Patel, M. H., *Phys. Rev. Letters* **70**, 1112-1115 (1993).
7. Luenberger, D. G., *Introduction to Linear and Nonlinear Programming*, Reading: Addison-Wesley, 1973.
8. Van Hove, M. A., Weinberg, W. H., and Chan, C. T., *Low Energy Electron Diffraction*, Berlin: Springer, 1986.
9. Jaynes, E. T. *Phys. Rev.* **106**, 620-630 (1957).
10. Collins, D. M., *Nature* **298**, 49-51 (1982).
11. Gilmore, C. J. Acta Cryst. A **52**, 561-589 (1996).
12. Saldin, D. K. Shneerson, V. L., and Wild, D. L., *J. Imaging. Sci. Technol.* **41**, 482-487 (1997).
13. Gerchberg, R. W., and Saxton, W. O., *Optik* **35**, 237-246 (1972)

On the Temperature Dependence of Multiple- and Single-Scattering Contributions in Magnetic EXAFS

H. Wende[1], F. Wilhelm[1], P. Poulopoulos[1], K. Baberschke[1],
J.W. Freeland[2]*, Y.U. Idzerda[2], A. Rogalev[3], D.L. Schlagel[4],
T.A. Lograsso[4], and D. Arvanitis[5]

[1] *Institut für Experimentalphysik, Freie Universität Berlin,
Arnimallee 14, D-14195 Berlin-Dahlem, Germany*
[2] *Naval Research Laboratory, Washington, D.C. 20375, USA,*
[3] *ESRF, B.P. 220, 38043 Grenoble, France,*
[4] *Ames Laboratory, Iowa State University, Ames, IA 50011, USA*
[5] *Department of Physics, Uppsala University, Box 530, 75121 Uppsala, Sweden*

Abstract. We demonstrate that the temperature dependence of structural as well as magnetic fluctuations can be probed by the use of the Magnetic Extended X-ray Absorption Fine Structure (MEXAFS) spectroscopy. We compare those to the dynamic disorder as probed by the EXAFS. Here we present temperature-dependent MEXAFS investigations carried out at the L-edges of a thin Fe film and a Gd single crystal. By comparing the experimental results to *ab initio* calculations the single-scattering contributions are separated from multiple-scattering contributions. It is found that the multiple-scattering contributions are enhanced for the MEXAFS compared to the normal EXAFS.

INTRODUCTION

Today, the Extended X-ray Absorption Fine Structure (EXAFS) spectroscopy is a standard tool to in order to investigate local structure and dynamics. The importance of its spin-dependent counterpart – the so-called Magnetic EXAFS (MEXAFS) – has been recognized after the first wiggles in the extended energy range have been detected in 1989 [1]. Now, adequate literature exists on the theoretical description of the MEXAFS phenomena [2–4]. In the present work we demonstrate that a basic understanding of the spin-dependent electron scattering can be achieved by comparing *ab initio* calculations carried out using the FEFF7 code [2] with experimental data. We find an enhancement of the multiple-scattering contributions for the magnetic EXAFS compared to the normal EXAFS. The separation of the multiple- from the single-scattering paths can be easily carried out by 'turning off' the contributions of the multiple-scattering paths in the calculations. The use of this computational tool assisted us to develop a better insight into the spin-dependent electron scattering. Although temperature-dependent EXAFS measurements have been shown to provide essential information on the local dynamics, it is surprising that only few MEXAFS measurements have been carried out as a function of temperature [5–9]. Here we present temperature dependent MEXAFS measurements at the $L_{3,2}$-edges of a polycrystalline Fe film and at the L_3-edge of a Gd single crystal. After the separation of the multiple- from single-scattering paths, the temperature dependence of those contributions can be studied individually.

EXPERIMENTAL DETAILS

The Gd single crystal was investigated at the ID12A beamline of the ESRF. The crystal had the shape of a plate and was magnetized normal to the surface (hard axis). The measurements were carried out at normal x-ray incidence using fluorescence yield. The MEXAFS measurements of the polycrystalline Fe film have been carried out at the NRL facility located at the U4B beamline of the NSLS. The film was magnetized in-plane (easy axis). The data were taken at 45° x-ray incidence in transmission geometry. The comparison of the normal EXAFS of the polycrystalline film to data recorded for an epitaxially grown Fe film on a Cu(100) substrate revealed that the polycrystalline film is well ordered on a local scale [8]. We have shown earlier that the overlap of the L_3- with the L_2-edge does not hinder the MEXAFS analysis [5–9]. The data for the Gd and the Fe samples were taken in an applied magnetic field high enough for magnetic saturation as was determined by means of element-specific hysteresis loops.

RESULTS AND DISCUSSION

We start with a qualitative discussion of the temperature dependence of the MEXAFS and EXAFS oscillations at the Gd L_3-edge. The experimental data are presented in Fig. 1 for two temperatures (10K and 250K). The upper panel shows the EXAFS wiggles which exhibit a clear temperature-dependent damping. Using a correlated Debye-model we determine a Debye-temperature of $\theta_D=155$K. This justifies the strong reduction of the EXAFS wiggles at 250K $> \theta_D$ since one expects an exponential attenuation via the EXAFS Debye-Waller factor $e^{-2\sigma^2 k^2}$. Here, σ^2 is the mean square relative displacement. The exponential damping is seen clearer when the intensity decrease at low k-values is compared to the damping at higher k-values. At $k=4.5$Å^{-1} the EXAFS signal at 250K is reduced to 69% of the 10K value whereas at $k=10.0$Å^{-1} the signal decreases to 16%. A similar damping of exponential form is found for the MEXAFS. This indicates that the thermal vibrations influence the magnetic EXAFS signal as well. Furthermore, there is at least an additional damping for the MEXAFS as the temperature increases due to the reduction of the magnetization.

We have observed such effect for Gd, Fe and Co [5-9]. This indicates that such effects are general in nature and are not linked to the specific electronic structure of the material. In order to discuss the different aspects of the temperature-dependent

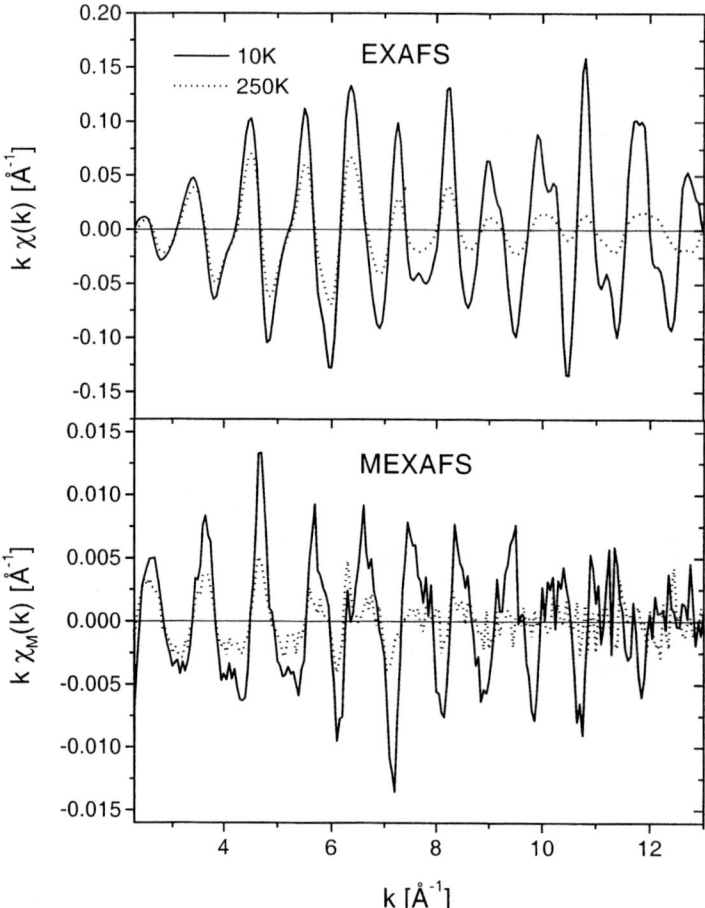

FIGURE 1. Temperature dependent Gd EXAFS $k\chi(k)$ and MEXAFS $k\chi_M(k)$ taken at the L_3 edge.

damping of the EXAFS and the MEXAFS in more detail, we now turn to a quantitative discussion for the EXAFS and MEXAFS of a polycrystalline Fe film. Before investigating the temperature dependence, it is useful analyze the Fourier transformed EXAFS and MEXAFS data at the lowest temperature. For a better understanding of the contributing scattering paths theoretical calculations were carried out using the *ab initio* FEFF7 code [2]. The results are presented in Fig. 2 where the Fourier transform of (a) the EXAFS and (b) the MEXAFS oscillations are shown. In both cases the dotted lines represent the results using single-scattering paths only. The calculations including single- and multiple-scattering contributions are given as solid lines. These calculations combining multiple- and single-scattering are in good agreement with the experimental data as can be seen in comparison to the inset of Fig. 3 and in Refs. [7,8]. The main peak of the EXAFS as well as the MEXAFS is determined by the nearest (1-2-1) and the next nearest (1-3-1) neighbor single-scattering paths. Comparing the calculation for the second and the third peak in the Fourier transform of the EXAFS and the MEXAFS a clear enhancement of the multiple-scattering contributions can be determined for the magnetic case. The triangular scattering path 1-2-3-1 contributes about 20% to the intensity of the second peak for the normal EXAFS whereas a contribution of about 40% is found for the MEXAFS. The enhanced multiple-scattering is also found for the third peak. Here numerous multiple-scattering paths are identified. The strongest contribution is due to the focusing path 1-2-6-2-1. For the third peak of the MEXAFS Fourier transform the multiple-scattering paths dominate the single-scattering ones, as about 65% of the peak intensity are due to multiple-scattering. A contribution of 50% only is found for the normal EXAFS. The enhancement of the multiple-scattering paths for the MEXAFS can be described in a phenomenological picture as discussed in Ref. [10]. The effect of the exchange interaction is introduced into the scattering process by means of a spin-dependent scattering amplitude F_M. This is scaled by the spin-polarization $\langle\sigma_z\rangle$ and is added to the Coulomb scattering amplitude F_0. Therefore, the backscattering amplitude becomes $F = F_0 \pm \langle\sigma_z\rangle \cdot F_M$ for right and left circularly polarized light, respectively. The backscattering amplitude for a multiple-scattering path of n scattering events can then be approximated to:

$$(F_0 + \langle\sigma_z\rangle \cdot F_M)^n \approx F_0^n \cdot \left(1 + n \cdot \langle\sigma_z\rangle \cdot \frac{F_M}{F_0}\right). \tag{1}$$

Thus, the multiple-scattering contributions can be enhanced for the MEXAFS by the factor n compared to the normal EXAFS.

Since the individual contributions to the Fourier transform peaks have been identified by means of the *ab initio* calculations, we now turn to the analysis of the experimental results. The temperature-dependent Fourier transforms of the experimental EXAFS and MEXAFS oscillations for the polycrystalline Fe film are presented in the inset of Fig. 3. A clear temperature-dependent damping can be seen for both cases. It was shown earlier that the lattice vibrations which determine the damping of the EXAFS are well described by a Debye-Temperature of θ_D=520K [7] analyzing the first EXAFS peak. An even stronger temperature-dependent damping is found for the first MEXAFS peak. We try now to correlate this damping to the magnetic part of the disorder. We show in Fig. 3 the reduced spontaneous magnetization as a function of the reduced temperature together with the intensity of the near edge Magnetic Circular X-ray Dichroism (MCXD) signal and (a) the first, (b) second and (c) third EXAFS and MEXAFS Fourier transform intensities. For

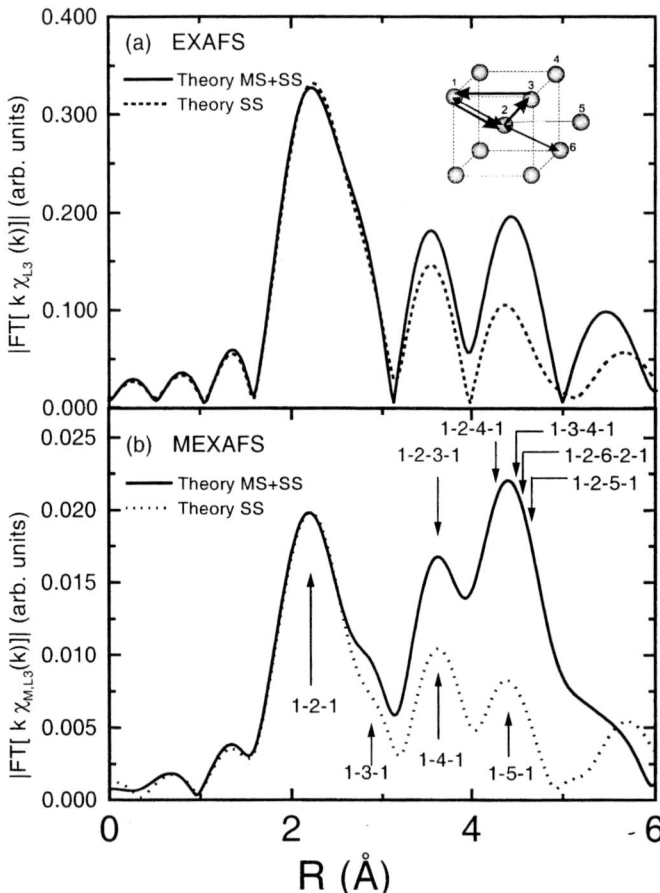

FIGURE 2. *Ab initio* calculation of Fourier transforms $|FT[k\chi_{L3}(k)]|$ and $|FT[k\chi_{M,L3}(k)]|$ of (a) the EXAFS (b) and MEXAFS oscillations for Fe bcc structure at 70K using the FEFF7 code. The single-scattering contributions (dotted lines) are separated from the combined multiple- and single-scattering contributions (solid lines). The peaks are assigned to the different scattering paths which are labeled according to the inset.

a relative comparison, the MCXD as well as the EXAFS and MEXAFS intensities are scaled to match the literature at the lowest temperature. The MCXD signal follows the temperature dependence of the magnetization as shown in Fig. 3(a).

This is not the case for the first EXAFS and MEXAFS peak of the Fourier transform. First we discuss why the EXAFS peak exhibits a much stronger temperature dependence. The temperature dependence of the normal EXAFS is determined by the Debye-temperature of $\theta_D=520$K. Therefore a much stronger damping is

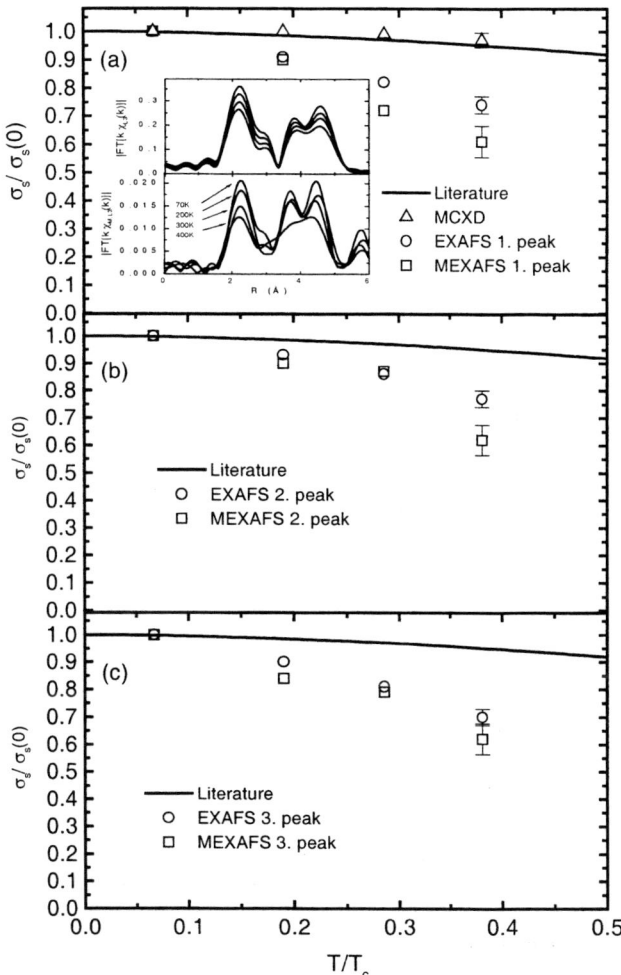

FIGURE 3. Temperature dependence of the EXAFS and MEXAFS Fourier transform intensities for the polycrystalline Fe film. The reduced spontaneous magnetization is given as a function of the reduced temperature (taken from the literature [11]). The Fourier transform intensities are scaled to match the literature values at the lowest temperature. The inset shows the temperature dependence of the Fourier transforms for the experimental EXAFS (top) and MEXAFS (bottom) oscillations.

found at T=400K (which corresponds to a reduced temperature of T/T_C=0.38) in comparison to the reduction of the MCXD signal which is determined by the Curie-temperature of T_C=1050K. This is due to the fact that up to a reduced temperature of $T/T_C \approx 0.3$ the attenuation of the MCXD signal is described by a $1 - \beta T^{\frac{3}{2}}$ law. This shows a much smaller temperature-dependent decrease than the exponential damping which describes the normal EXAFS. A surprising result is therefore the even stronger temperature dependence of the first MEXAFS Fourier peak in comparison to the EXAFS signal (Fig. 3(a)). At a temperature of 400K (T/T_C=0.38) the MCXD signal is reduced to only 97% of the T=0K value whereas the EXAFS signal is reduced to 74%. This shows that a simple multiplication of those values leading to 72% does not describe the observed damping of the MEXAFS of 61% with respect to the T=0K value (Fig. 3). This indicates that there is a larger magnetic disorder on a local scale (probed with the MEXAFS) compared to the long-range spin fluctuations (probed with MCXD) leading to the decrease of the magnetization. This difference in the "probing length" of the MCXD and the MEXAFS technique is due to the fact, that the mean free path of the scattered photoelectron is a function of the kinetic energy of the electron and differs for both cases. In the near-edge energy range (MCXD) the mean free path tends to diverge and therefore a long-range order is probed. In contrast, the MEXAFS probes the magnetism on a nearest neighbor length scale due to its more localized scattering origin and the fact that the mean free path exhibits a minimum of about 7Å at k=3.0Å^{-1} and increases to about 30Å at k=13.0Å^{-1}. Up to now we discussed the first peak of the Fourier transform which includes single-scattering contributions only. The second and third Fourier transform peaks contain strong multiple-scattering contributions for the MEXAFS case as can be seen in Fig. 2. This can be the origin of the observed anomaly in the damping of the MEXAFS signals around T/T_C=0.18, where a slower intensity decrease is observed (see Fig. 3 (b) and (c)). It is known that multiple-scattering contributions exhibit a stronger temperature-dependent damping. Therefore the damping of the multiple scattering paths will be seen mostly at lower temperatures, whereas the damping of the single-scattering paths will become effective at higher temperatures. Indeed for the second as well as the third peaks the same trend of a stronger temperature-dependent damping of the MEXAFS in comparison to the EXAFS is found in Fig. 3 (b) and (c) around T/T_C=0.4. These observations allow us to set boundaries to the temperature range in which each of those two mechanisms is mostly responsible for the damping.

CONCLUSION

We have presented two temperature-dependent magnetic EXAFS studies. For the Gd single crystal as well as for the polycrystalline Fe a clear temperature-dependent damping for the normal EXAFS and the magnetic EXAFS were determined. The quantitative analysis for the polycrystalline Fe film shows that thermal vibrations also influence the MEXAFS signal. The comparison of the damping of the normal EXAFS with the MEXAFS demonstrates that there must be a higher local magnetic disorder compared to the long-range order probed with MCXD. *Ab initio* calculations carried out with and without multiple-scattering contributions clearly indicate an enhancement of multiple-scattering paths for the magnetic EXAFS. This separation enabled us to discuss the temperature dependence of the individual multiple- and single-scattering paths.

ACKNOWLEDGMENTS

This work is supported by BMBF (05 SC8KEA-3), DFG (Sfb 290) and ESRF (experiment HE-536). We want to thank J.J. Rehr and A.L. Ankudinov for helpful discussions and providing the most recent FEFF versions to us.

REFERENCES

*Permanent address: Adv. Photon Source, Argonne, IL 60439, USA

1. Schütz, G., Frahm, R., Mautner, P., Wienke, R., Wagner, W., Wilhelm, W., and Kienle, P., *Phys. Rev. Lett.* **62**, 2620 (1989).
2. Ankudinov, A.L., and Rehr, J.J., *Phys. Rev. B* **56**, R1712 (1997).
3. Brouder, Ch., Alouani, M., Bennemann, K.H., *Phys. Rev. B* **54**, 7334 (1996).
4. Ebert, H., Popescu, V., Ahlers, D., Schütz, G., Lemke, L., Wende, H., Srivastava, P., and Baberschke, K., *Europhys. Lett.* **42**, 295 (1998).
5. Lemke, L., Wende, H., Srivastava, P., Chauvistré, R., Haack, N., Baberschke, K., Hunter-Dunn, J., Arvanitis, D., Mårtensson, N., Ankudinov, A., and Rehr, J.J., *J. Phys.: Condens. Matter* **10**, 1917 (1998).
6. Srivastava, P., Lemke, L., Wende, H., Chauvistré, R., Haack, N., Baberschke, K., Hunter-Dunn, J., Arvanitis, D., Mårtensson, N., Ankudinov, A., Rehr, J.J., *J. Appl. Phys.* **83**, 7025 (1998)
7. Wende, H., Freeland, J.W., Chakarian, V.,Idzerda, Y.U., Lemke, L., and Baberschke, K., *J. Appl. Phys.* **83**, 7028 (1998).
8. Wende, H., Srivastava, P., Arvanitis, D., Wilhelm, F., Lemke, L., Ankudinov, A., Rehr, J.J., Freeland, J.W., Idzerda, Y.U., and Baberschke, K., *J. Synchrotron Rad.* **6**, 696 (1999).
9. Wende, H., Ph.D. thesis, Freie Universität Berlin, *Extended X-ray Absorption Fine Structure Spectroscopy of Surfaces and Thin Films: Local Structure, Dynamic and Magnetic Properties*, Berlin: Verlag Dr. Köster (1999), ISBN 3-89574-341-0
10. Schütz, G., and Ahlers, D., in *Lecture Notes in Physics: Spin-Orbit-Influenced Spectroscopies of Magnetic Solids*, edited by H. Ebert and G. Schütz, Berlin: Springer (1996), p. 229
11. Stearns, M.B., in *Magnetic Properties of Metals*, edited by Wijn H.P.J., Landolt-Börnstein, Vol. III/19a, Berlin: Springer (1986), p. 37

EXAFS and thermal expansion

G. Dalba*, P. Fornasini*[†], R. Grisenti*, and F. Rocca[‡]

*Dipartimento di Fisica dell'Università degli Studi di Trento
and Istituto Nazionale di Fisica della Materia, I-38050 Povo (Trento), Italy
[‡]CEFSA - Centro CNR-ITC di Fisica degli Stati Aggregati, I-38050 Povo (Trento), Italy
[†]e-mail: fornasin@science.unitn.it

Abstract. The sensitivity of EXAFS to thermal expansion has been experimentally studied on several crystals: Ge, CdSe, and AgI. In no case does the first cumulant reproduce the thermal expansion, owing to relative atomic vibrations normal to the bond. By converse, EXAFS can give original information on the average perpendicular relative displacement $\langle \Delta u_\perp^2 \rangle$. The thermal expansion of germanium can be determined from the 3rd cumulant, provided that quantum effects are taken into account. For CdSe and AgI, on the contrary, the 3rd cumulant does not reproduces the thermal expansion.

I INTRODUCTION

The basic property which makes EXAFS a powerful structural tool is the sensitivity to the distance between absorber and backscatterer atoms. However, to establish an accurate relation between the real interatomic distances in a solid and the parameters which are usually obtained from EXAFS analyses is far from trivial. The main difficulties come from the reduction of two tri-dimensional distributions of atomic positions (described by thermal ellipsoids, whose centers define the interatomic distance) to a one-dimensional distribution of distances, which is in turn sampled by the photoelectron spherical wave. In this process, the correlation between atomic thermal motions plays a fundamental role [1,2]. However, while the relative displacement parallel to the bond direction $\langle \Delta u_\parallel^2 \rangle$ can again be obtained from EXAFS data, the relative displacement normal to the bond direction $\langle \Delta u_\perp^2 \rangle$ has to be independently evaluated from vibrational calculations. An accurate relation between EXAFS parameters and interatomic distances is important for testing calculated amplitudes and phaseshifts against experimental reference systems or for evaluating the effective length of multiple scattering paths. It is also useful when temperature dependent measurements are performed with the aim of determining the local thermal expansion.

In this paper we present and compare experimental results obtained by measuring EXAFS as a function of temperature on several crystals characterized by different structure and thermal properties: Ge [3], CdSe [4] and AgI [1]. To maximize the amount of information directly available from experimental spectra and minimize theoretical biases, the data analysis was limited to the first shell and carried on by the cumulant method within the single scattering approximation, taking the lowest temperature spectrum as reference for backscattering amplitudes, phaseshifts and anelastic terms. By this procedure only relative values of the cumulants ΔC_i with respect to the lowest temperature spectrum were obtained; the attention was then focussed on thermal expansion, rather than on absolute values of distance. Two different procedures for measuring thermal expansion were attempted, based on the temperature dependence of the *first* and *third* cumulant, respectively.

II RESULTS

One possibility for measuring thermal expansion relies on the study of the temperature dependence of the *1st cumulant*, say the mean value of the distribution of distances. The first cumulants C_1 and C_1^* of the effective and real distributions [5], respectively, are connected to the distance R between the centers of thermal ellipsoids through the equations [4]

$$C_1 = C_1^* - \frac{2 C_2^*}{C_1^*}\left(1 + \frac{C_1^*}{\lambda}\right); \qquad C_1^* = R + \frac{C_\perp}{2R} \qquad (1)$$

where λ is the photoelectron mean free path, and $C_\perp = \langle \Delta u_\perp^2 \rangle$ is the *perpendicular* component of the average relative thermal displacement: $\langle \Delta u_\perp^2 \rangle = \langle (\vec{u}_j - \vec{u}_0)^2 \rangle - \langle \Delta u_\parallel^2 \rangle$. The *parallel* component corresponds to a good approximation to the second cumulant: $\langle \Delta u_\parallel^2 \rangle \simeq C_2^* \simeq C_2$. (The differences between cumulants of the real and effective distributions of second and higher orders are usually neglected).

The first of eqs. (1) can be inverted using the experimental values $\Delta C_2^* \simeq \Delta C_2$ to get a workable relation between relative first cumulants of the effective and real distributions [3]. In Fig.1 the relative values ΔC_1 and ΔC_1^* are compared with the thermal expansion ΔR, taken from refs. [6,7]. As expected from the first of eqs. (1), ΔC_1^* grows faster than ΔC_1 for all the crystals investigated. Besides, both ΔC_1^* and ΔC_1 grow faster than ΔR. The difference between ΔC_1^* and ΔR, due to the relative perpendicular displacements $C_\perp$, represents a non negligible *apparent thermal expansion*. The true thermal expansion ΔR can then be recovered from an EXAFS experiment only if the term $C_\perp$ is known in advance from independent vibrational calculations.

Conversely, if the thermal expansion is known, the EXAFS results can be exploited to get original information on vibrational properties. The perpendicular relative displacement term $C_\perp = \langle \Delta u_\perp^2 \rangle$ can be obtained by fitting an Einstein model to the relative values $\Delta C_\perp$ calculated from ΔC_1^* and ΔR by inverting the second of eqs. (1). Like the parallel relative displacement term $C_2 = \langle \Delta u_\parallel^2 \rangle$, also the perpendicular term $C_\perp$ is an independent test of eigenfrequencies and eigenvectors from vibrational models or ab-initio calculations. A useful parameter for

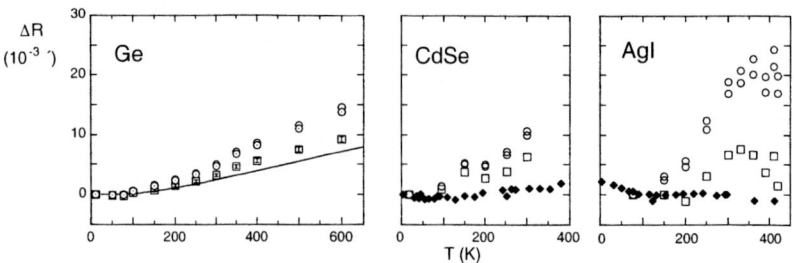

FIGURE 1. Relative first cumulants ΔC_1 (squares) and ΔC_1^* (circles) of 1st shell effective and real distributions, respectively, for Ge, CdSe and AgI (the experimental error bars, here omitted for clarity, can be found in the original references [3,4,1]); upper and lower circles for ΔC_1^* correspond to $\lambda = 6$ or 12 Å, respectively. The continuous line for Ge and the diamonds for CdSe and AgI represent thermal expansion ΔR from Ref. [6] (and for AgI also from Ref. [7].)

comparing theory with experiment is the ratio between the two terms, $\gamma = C_\perp/C_2$, which should be nearly constant above the Debye temperature θ_D. For germanium the ratio stabilizes at about $\gamma \simeq 6$ above θ_D [3], and its overall behavior is in good agreement with the same ratio calculated for silicon using an adiabatic bond charge model [8]. For CdSe and AgI the ratio stabilizes at higher values ($\gamma \simeq 10$) above θ_D. The difference between Ge on the one hand and CdSe and AgI on the other can be attributed to the stronger correlation of vibrational motion along the bond direction in wurtzite structures [9], which reduces C_2 with respect to $C_\perp$.

An alternative possibility for measuring thermal expansion from EXAFS is based on the *third cumulant*, which is connected to the asymmetry of the distribution of distances. Classically the temperature dependence of the third cumulant is, to first order, proportional to T^2 [10], and thermal expansion is often calculated as $C_3/2C_2$ [11]. Recent quantum statistical calculations have shown that quantum effects can be not negligible, giving rise to a non-zero value at 0 K [12-14]; an expression connecting thermal expansion to EXAFS cumulants has been explicitly derived by Frenkel and Rehr [12].

In systems with relatively low θ_D, like AgI and CdSe, the experimental values ΔC_3 were consistent with a T^2 behavior; as a consequence, low temperature quantum effects were considered negligible and the absolute value of C_3 could be assumed equal to zero at zero Kelvin, so that $C_3 = \Delta C_3$ [1,4]. In the case of germanium, on the contrary, the experimental ΔC_3 values were characterized by a flat region at low temperatures, which was attributed to a quantum deviation from the classical T^2 approximation [3].

The ratios $C_3/2C_2$ for Ge, CdSe and AgI are compared with thermal expansion ΔR in Fig. 2.

In the case of germanium, if quantum effects in C_3 are properly taken into account, the ratio $C_3/2C_2$ reproduces the thermal expansion, and is equivalent to the net thermal expansion $a = -3k_3C_2/k$ as defined by Frenkel and Rehr [12], in spite

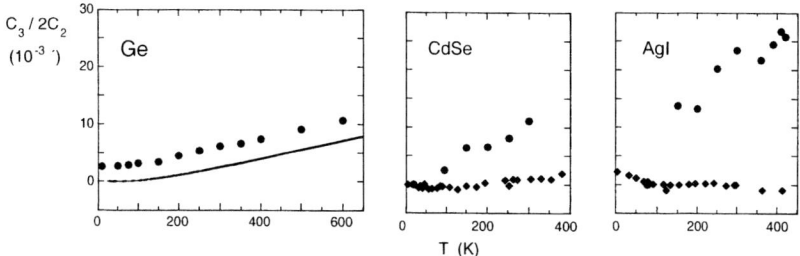

FIGURE 2. Comparison between the ratio $C_3/2C_2$ (full circles) and thermal expansion ΔR (continuous line for Ge, diamonds for CdSe and AgI.)

of the difference of analytical expressions [3]. The difference in absolute values of about 3×10^{-3} Å is due to the shift of the equilibrium position at 0 K with respect to the minimum of the effective potential. If quantum effects are neglected, and the classical approximation is assumed valid for C_3 of germanium, the values $C_3/2C_2$ slightly overestimate the thermal expansion [3].

In the case of CdSe and AgI the ratios $C_3/2C_2$ largely overestimate the thermal expansion, which is actually very weak for CdSe and almost null for AgI. For these two crystals, then, the thermal expansion cannot be obtained nor from the first nor from the third cumulant.

III DISCUSSION AND CONCLUSIONS

In the previous section the thermal expansion has been compared with the temperature dependence of two EXAFS parameters: the first cumulant C_1^* of the real distribution and the third cumulant $C_3 \simeq C_3^*$. For all systems considered the first cumulant does not reproduce the thermal expansion, owing to the effect of thermal vibrations normal to the bond direction (Fig. 1). The third cumulant reproduces the thermal expansion of germanium, but in the case of CdSe and AgI is in complete disagreement (Fig. 2).

The distributions of distances at different temperatures can be connected to an effective pair potential. If this effective potential were temperature independent, then the mean values of the distributions of distances would depend only on the potential anharmonicity, and the first cumulant should carry the same information as the third one. The observed lack of equivalence between first and third cumulants can be explained by assuming that the minimum position of the effective potential is temperature dependent. The growth of the mean value C_1^* with temperature is then a joint minimum position and the anharmonicity of the potential.

The results obtained for germanium suggest that the anharmonicity of the *effective potential* reflects essentially the anharmonicity of the *crystal potential*, so that the third cumulant actually gives the thermal expansion. The thermal vibrations normal to the bond direction produce instead a positive shift of the minimum of the effective potential, which causes the apparent thermal expansion of the first cumulant, without strongly affecting the shape of the potential. Actually, if the distributions of distances are reconstructed from EXAFS cumulants at various temperatures, the shift of their maximum positions corresponds to the difference between ΔC_1^* and the true thermal expansion ΔR.

The situation is more complicated for CdSe and AgI. Here the distributions of distances reconstructed from EXAFS cumulants exhibit a strong negative shift with temperature of their maximum positions, probably connected with the anomalous thermal expansion of these compounds. In the absence of an independent knowledge of the term $C_\perp$, it is impossible to disentangle these three effects (downward shift of the potential, thermal vibrations normal to the bond direction, anharmonicity) from the measurements of only two parameters (first and third cumulants).

In conclusion, the accurate evaluation of thermal expansion of crystals from EXAFS experiments depends on their vibrational properties. This dependence is different for different crystals. In any case it affects also the determination of absolute values of distance. A further study of these effects will help in assessing the limits of accuracy of EXAFS interpretations, based either on phenomenological analyses or on theoretical simulations. Besides, new possibilities are open for using EXAFS as a probe of local vibrational properties.

REFERENCES

1. G. Dalba, P. Fornasini, R. Gotter, and F. Rocca, Phys. Rev. B **52**, 149 (1995).
2. E.A. Stern, Journal de Physique IV, **7**, C2, 137 (1997).
3. G. Dalba, P. Fornasini, R. Grisenti, and J. Purans, Phys. Rev. Lett, **82**, 4240 (1999).
4. G. Dalba, P. Fornasini, R. Grisenti, D. Pasqualini, D. Diop, and F. Monti, Phys. Rev. B **58**, 4793 (1998).
5. E.D. Crozier, J.J. Rehr and R. Ingalls, in *X-ray Absorption*, edited by D.C. Koningsberger and R. Prins (Wiley, New York, 1988).
6. Y.S. Touloukian, R.K. Kirby, R.E. Taylor, and P.D. Desai, *Thermophysical properties of matter* (Plenum, New York, 1977), Vol. 13.
7. A. Yoshiasa et al., Acta Crystallogr. B **43**, 434 (1987).
8. O.H. Nielsen and W. Weber, J. Phys. C **13**, 2449 (1980).
9. G. Dalba, P. Fornasini, F. Rocca, and S. Mobilio, Phys. Rev. B **41**, 9668 (1990).
10. J.M. Tranquada and R. Ingalls, Phys. Rev. B **28**, 3520 (1983).
11. L. Tröger, T. Yokoyama, D. Arvanitis, T. Lederer, M. Tischer, and K. Baberschke, Phys. Rev. B **49**, 888 (1994).
12. A.I. Frenkel and J.J. Rehr, Phys. Rev. B **48**, 585 (1993).
13. T. Fujikawa and T. Miyanaga, J. Phys. Soc. Jpn. **62**, 4108 (1993).
14. T. Miyanaga and T. Fujikawa, J. Phys. Soc. Jpn. **63**, 1036 (1994).

Dynamical scattering of X-rays by real binary crystals and problem of point defects

L.I.Datsenko[*], V.P.Klad`ko [*], V.F.Machulin [*], S.Manninen [$],
I.V.Prokopenko[*]

Institute of Semiconductor Physics NASU, Kiev, Ukraine
[$]*Department of Physics, Helsinki University, Helsinki, Finland*

Abstract. One of the important subjects in the material science of semiconductor compounds is a problem of stoichiometry. The methods based on the measurement of kinematical integral reflectivity (IR), R_i^K, for the quasi-forbidden reflection (QFR) of X-rays were formerly used for investigation of the $GaAs$ composition. To determine the value of deviation from stoichiometry, $\Delta = c_A - c_B$ where c_i is a concentration of the component A or B, one should make some corrections for extinction phenomena which is difficult to take into account for a real (with defects) crystal. The reflectivity of a real crystal for a QFR, R_i^D, may be also described within the dynamical theory taking into account the Debye-Waller static factor, L_H, the extinction coefficient μ_{ds}, and parameter Δ. To determine these characteristics the experimental thickness pendulum oscillation of $R_i^D(t)$ (200 reflection, λ=0.1198nm) was measured and analysed for the first time. By this the Hönl corrections of atomic formfactors for anomalous dispersion were made. Another independent approach consisted in analysis of the experimental energetic dependence of the $R_i(\lambda)$ for the wavelengths situated between the two absorption K-edges was used too. Relatively close values of the L_H and μ_{ds} for GaAs crystals with dislocations as well as parameter Δ were obtained by fitting of the $R_i^D(t)$ and $R_i^D(\lambda)$ nonlinear functions, calculated by the theory, to the experimental data.

INTRODUCTION

Nonstoichiometry and point-like defects is one of the most important problems in the material science of binary crystals which are widely used today in solid state devices technology [1]. It is well known that these defects are usually investigated by means of various optical and electrophysical methods [2]. Results obtained by these methods are needed to be proved by other independent investigations, for example, by the x-ray diffractometrical ones . But they, unfortunately have very low sensitivity to points defects even in the case when such dynamical phenomena as Borrmann effect are used [3]. Diffuse scattering of X-rays is sensitive to defect clustering only [4]. So for an enhancement of X-ray of X-ray scattering one use sometimes laborious methods [5]. Nevertheless utilisation of reflectivities of so called quasiforbidden reflection permits to study nonstoichiometrical distortions of a binary crystal sublattice [6]. Such approach supposes utilization kinematical case of diffraction where the total reflectivity does not depend on structure perfection of a crystal. More realistic situation is however the case of dynamical scattering when a real defect structure should be taken into account [7].

The aim of this paper was to obtain the quantitative independent information not about nonstoichiometry of a GaAs crystal only but the values of integral characteristics of crystal perfection i.e. the Debye-Waller static factor L_H and coefficient of extinction due to diffuse scattering on defects μ_d too. For this purpose the measurements of the Pendellösung intensity oscillations for the QFRs and so called energetical or wavelength dependent reflectivities were carried out. These experiments were made for the wavelengths of X-ray continuous spectrum situated close to the absorption K-edges of the As and Ga components where the Hönl corrections for atomic formfactors due to anomalous dispersion are essential what is of principle for developed experimental methods.

THEORETICAL BASES OF EXPERIMENTAL METHOD

Thickness oscillations of reflectivities for the QFR`s

Pendellösung fringes in a differential reflectivity R as it is well known may be described for a perfect crystal by the following formula accounting contribution of the both Bloch`s waves with a weak and strong absorption as well as oscillation term :

$$R = 0.25 \cdot \left\{ \exp\left[-\frac{\mu t}{\cos\vartheta}(1-\varepsilon)\right] + \exp\left[-\frac{\mu t}{\cos\vartheta}(1+\varepsilon)\right] - 2\exp\left[-\frac{\mu t}{\cos\vartheta}\right]\cos(2\alpha) \right\} \quad (1)$$

were $\alpha = \pi |\chi_{rh}| ct / (\cos\vartheta \cdot \lambda)$ is the ratio of thickness t and extinction length Λ. Here μ, ε, χ_{rh}, C stand for coefficient of absorption, parameter of a wave field localization in a crystalline lattice, real part of the Fourier coefficient of susceptibility and polarisation factor. χ_{rh} as well as imaginary part of the mentioned Fourier coefficient χ_{ih} depend on the corresponding part of structure factor F_h :

$$\chi_{rh} = r_a (\frac{\lambda^2}{\pi V}) F_{rh}$$

$$\chi_{ih} = r_a (\frac{\lambda^2}{\pi V}) F_{ih} \quad (2)$$

where r_a, λ and V are respectively the classical electron radius ; wave– length of radiation and a volume of an elementary lattice . For a QFR the F_h value is proportional to the difference of atom form factors of components :

$$F_H = 4 (F_{Ga} C_{Ga} 2057 - F_{As} C_{As}) \qquad (3)$$

where C_{Ga} and C_{As} are the concentrations of components in corresponding sublattices. By measurements using the wavelengths situated near the absorption K-edges the Hönl corrections for the real $\Delta f\,'$ and imaginary $\Delta f\,''$ part of scattering should be taken into account:

$$f = f_0 + \Delta f' + i\Delta f'' \qquad (4)$$

When some disorder there is in one of sublattices, let in the A, the parameter nonstoichiometry $\Delta = C_A - C_B$ may be introduced. So the QFR's are perspective for determination of the Δ parameter which is determined by a little variation of a component concentration. For the case of substitution of native atoms by the others ones (let it be the silicon atoms with concentrations C_{Si}) the structure factor of such crystal may be written in the following way:

$$F_h = \{(1 - C_{Si})[(f_0 + \Delta f')_{As} + \Delta f''_{As}] + C_{Si}(f_0 + \Delta f')_{Si} - C_{Ga}[(f_0 + \Delta f')_{Ga} + \Delta f''_{Ga}]\} \qquad (5)$$

So when measuring the Pendellösung fringes distance for the QFRs one may determine the parameter Δ contrary the case of an usual structural reflection where it can not be done. But practically this procedure is difficult to realize because the extinction distance Λ for the QFRs is very large due to little value of the corresponding structure factor F_h (3). This obstacle can be met using the wavelengths situated near the Ga absorption K-edge [7]. In this case the Hönl corrections Δf_{Ga} may considerably change the value of F_h. All the said relates to the case of a perfect crystal. For a real crystal with structure defects the integral reflectivity R_i for the QFRs depends on the Debye-Waller factor L_H and the coefficient of additional energy losses parameter μ_{ds} due to diffuse scattering on defects (extinction). In this case the formulas of the Molodkin dynamical theory [8] should be used taking into account the Bragg, R_B, and diffuse, R_D, components of total reflectivity R_i:

$$R_i = R_B + R_D \,. \qquad (6)$$

Using the mentioned formulas and the known fitting procedure for the measured thickness dependencies of a total reflectivity, R_i^{exp}, one can determine not only the structure perfection characteristics L_H and μ_d of a real crystal but the nonstoichiometry parameter Δ too [7].

Energetical dependence of the R_i for the wavelengths situated between the Ga and As absorption K-edges.

Utilization of the continuous spectra of X-rays gives another (independent) possibility for determination of the mentioned three parameters. In this region of wavelengths a special point exists where the real part of a structure factor and therefore the corresponding value of χ_h go to zero due to effect of the Hönl correction

[9,10].. In this case we were trying to apply again the Molodkin`s theory developed for the Bragg case of diffraction [11,12]. So both of the components R_B and R_D of a total reflectivity R_i in (6) now depend on the L_H and μ_d parameters :

$$R_B(\Delta\vartheta) = \xi(L - \sqrt{L^2 - 1}) \qquad (7a)$$

$$R_D(\Delta\vartheta) = F_{din}(\Delta\vartheta)\frac{\mu_d(k_0)\gamma_0}{2\mu_i(\Delta\vartheta)}. \qquad (7b)$$

Meanings of the diffraction parameters ξ and L are denoted in [13] . The values $F_{dyn}(\Delta\theta)$, $\mu_i(\Delta\theta)$, $\mu_d(k_0)$ were discussed in [14]. One should note by this that the coherent component of scattering (7a) depends on the Hönl correction of the corresponding atom form factors F_{Ga} and F_{As} for anomalous scattering near the absorption K-edges (4). It permits to calculate the reflectivity of a crystal near the mentioned specific point where the F_{rh} and $\chi_{rh} \to 0$.

PECULIARITIES OF THE EXPERIMENTAL TECHNIQUES

The thicknes Pendellösung fringes were observed by us for the first time for 200 reflections of a *GaAs* crystal using the wavelength $\lambda=1.1965 Å$ situated close to the Ga absorption K-edge ($\lambda_{KGa}=1,1957 Å$) in the longwave region[7]. Supply regime of the X-ray unit (U= 20 kV,I = 30 mA) permitted us to get rid of multiple harmonics of continuous spectrum. Very thin (t~100µm) samples were used to get an intensity level to be in excess of the background .They were step-like tilted in the angular interval $\alpha = \pm 60°$ around the diffraction vector direction. The normalized values of reflectivity r :

$$r = \overline{R}_i / R_i^K \cdot 2A + 1 - I_0(\mu t \varepsilon) \qquad (8)$$

were used to exclude an influence of photoelectric absorption. Here R_i and R_{iK} are respectively the measured value and calculated one for the kinematical case of diffraction. $A = \pi t/\Lambda$. $I_0(\mu\varepsilon t)$ is the Bessel function of zero order. For the wavelength $\lambda = 1,1965 Å$ the extinction length Λ_{200} is equal to 0,0053cm, where as for the 400 reflection this value is equal to 0,0016cm. To illustrate the difference of Λ values between the QFR 222 and the usual reflection 111 the calculations according formula (1) were carried out for a GaP crystal (Fig.1). One can see the larger Pendelösung distance for 222 reflection even in the case of a GaP where a difference $f_{Ga} - f_{As}$ is not so little as in GaAs sample. So utilization of the longwave region near the Ga absorption K-edges is justified The similar methods were used for measurements of the energetical dependences of reflectivities for the wavelengths in the interval between λ_{KGa} and λ_{Kas} excepting only the tilt of a sample. By calculation of reflectivities close to the point where the imaginary part of F_h ,i.e. F_{ih} exits only this last value was taken into account. The parameters of structure perfection i.e. L_H, μ_d and Δ have been determined by the fitting procedure of the calculated value $R_i^T(\lambda)$ to the experimental ones $R_i^{exp}(\lambda)$. The minimum of the functional:

$$\Phi = \sum_{i=1}^{n}\left[R_i^{exp}(\lambda)_i - R_i^T(\lambda)_i\right]^2 / \sigma_i^2 \qquad (9)$$

was looking for. The number of iterations was chosen in such a way that the accuracy of calculation was not worse $2 \div 3$ %.

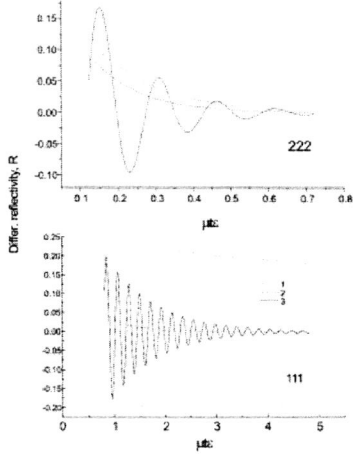

Fig.1. Calculation of the Pendellösung phenomenon for the 222 QFR and "usual" 111 reflection for a GaP perfect crystal (formula 1). Dashed and dotted lines correspond to the Bloch waves with low and strong absorption of X-ray (λ= 1.1984 Å

Results and their discussions

The thickness dependence of the normalized reflectivity of the 200 QFR r to a kinematical value R_i^K is shown in the (Fig 2) for the *GaAs* crystal containing some dislocation (density $N_d = 8 \bullet 10^4 cm^{-2}$). One can see difference between the real nonstoichiometrical sample (the curve 2) and for a perfect stoichiometric crystal (curve 1). The fitted calculated curve is shown by the solid line in the upper part of the graph. One can see also that the upper graph (measured and calculated) are considerably displaced relatively lower curve 1 due to effect of diffuse scattering parameter μ_d. The curve 2 changes also the maxima and minima coordinates . It is an effect of structure defects (dislocations) via parameter of L_H. The last effect depends also on the level of nonstoichiometry because the best fit (the solid line 2 in the upper graph) can be reached in the case only when the parameter Δ is taken into account.

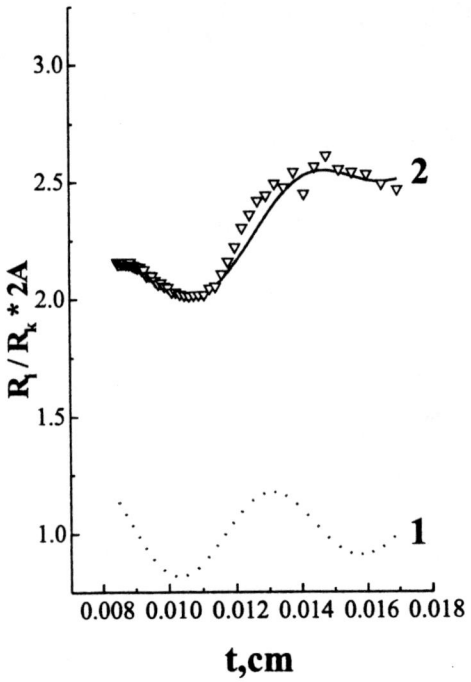

Fig.2. Thickness dependences of a reflectivity for the GaAs crystal containing $8 \cdot 10^4 \text{cm}^{-2}$ dislocations (curve 2) and for a perfect stoichiometric sample (calculation) (curve 1). $\lambda = 1.1984$ Å. The experimental results are shown by points.

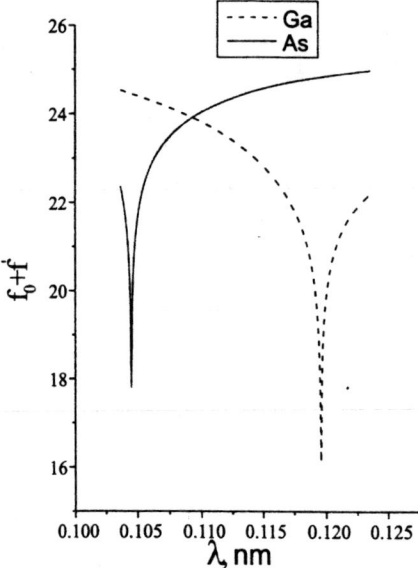

Fig.3. Variations of the Ga and As formfactors with changing of wavelengths between λ_{KGa} and λ_{KAs}.

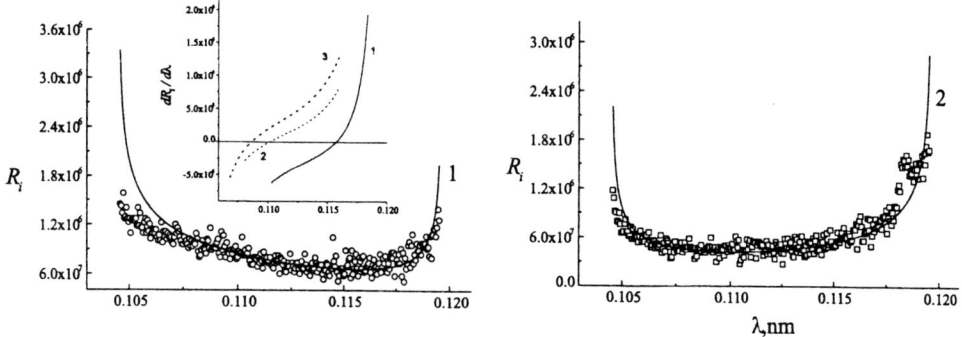

Fig.4. The energetical dependences of a reflectivity for GaAs crystals dopped with Si. C_{Si} are equal 1

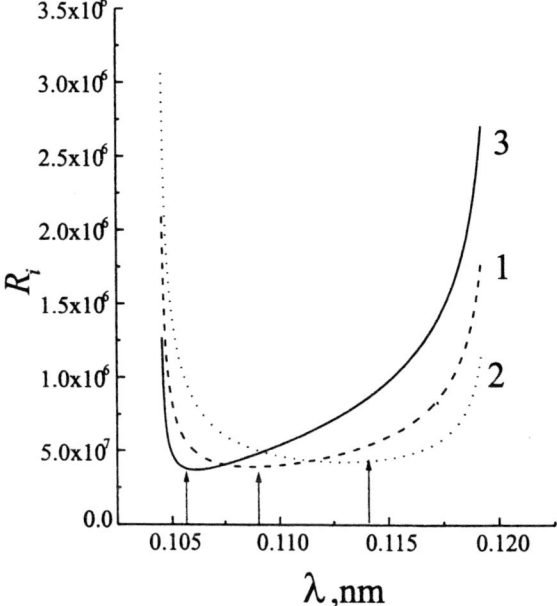

Fig.5. Effect of Ga(2) or As(3) excess on the reflectivity minimum close to the point where $F_{rh} \to 0$. The stoichiometrical sample is shown by curve 1.

Let us proceed to the results of investigations of the energetical dependences of R_i. First of all we consider the character of gallium and arsenic atom formfactors variations (Fig.3.) One can see that there is an equality of $(f_0 + \Delta f')_{Ga}$ and $(f_0 + \Delta f')_{As}$ values for the one specific wave length where $F_{rh} \to 0$. The exact positions of such points which are directly connected with a level of nonstoichiometry of a crystal can be determined by the known procedure of an extremum searching. Behaviour of the $dR_i/d\lambda$ function is shown in the insert to the Fig.4 for the fitted curves 1 and 2 for the samples with different concentration of Si atoms in GaAs crystals.

Effect of the Ga and As excess on the position of R_i minimum (pointers) is shown in the calculated curves 2 and 3 relatively the the stoichiometric sample (Fig.5.). So when analyzing the shape of the $R_i(\lambda)$ energetical dependences one can judge about the excess of one or other component even on a qualitative level. The carried out investigations have shown that structure defects do not influence the position of the discussed point. Really crystal imperfections can only displace the corresponding curves along the y axis.

Let us now compare the results of the L_H, μ_d, and Δ parameters determination by two discussed methods (Table 1). It is ease to see that all of these parameters are relatively close. This correlation can be considered as satisfactory taking into account that it is difficult to measure the same point on the surface of a sample.

rdrrMethod	L_H	μ_d, cm^{-1}	Δ
Pendellösung fringes	0.052 ± 0.004	31 ± 3	0.00030
Energetical dependences of R_i	0.067 ± 0.003	18 ± 3	0.00032

Table 1. Integral characteristics of structure perfection, L_H, μ_d and parameter of nonstoichiometry for GaAs crystal as determined by the two independent methods for the 200 qusiforbidden reflection

CONCLUSIONS

Possibility of the integral characteristics of structure perfection (L_H, μ_d) and nonstoichiometry parameter Δ determination by the two independent experimental methods based on the integral reflectivity measurements for a quasiforbidden reflection of X-ray continuous spectrum was shown. Contrary to the known methods used the kinematical approximation of scattering, utilization of the dynamical phenomena (Pendellösung fringes or energetical dependencies of a reflectivity for the wavelengths situated between the absorption K-edges of binary crystal components) permits to study real binary crystals containing various structure defects. The developed methods is supposed to take into account the anomalous scattering phenomena near the K-edges of absorption of lattice components. The method of energetical dependence of reflectivity was developed for the Bragg case of diffraction what gives the unique possibility to study nondestructively the thin crystal film structures. This method uses the nonlinear character of energetical dependence of reflectivity with the special point were the imaginary part of structure factor exists only. Both of the methods assume utilization of the fitting procedure of calculated dependences by the Molodkin dynamical theory to experimental results and could be applied to experiments with synchrotron radiation.

REFERENCES

1. Bardsley W./ J.Cryst.Growth.-1980.-48. .-p.505
2. Lannoo M.,Bourgodin J. *Point defects in Semiconductors.* Berlin, Springer Verlang 1981, 263
3. Iveronova V.I., Katsnelson A.A., Kisin V.I., *Fizika Tvrdogo Tela* **11**, 3154 (1969).
4. Krivoglaz M.A. *Theory of x-ray and thermal neutron scattering by real crystals* 1969 New York; Plenum press p.192
5. Haubold H.-R, *G.Appl Cryst* **175** (1975)
6. Fujimoto I. Gap. *G. Appl Phys* **23** n5.-L287. (1984)
7. Datsenko L.J., KladkoV.P.,Melnyk V.M., Machulin V.F., *Metal physic and advanced technology* (in Russian). – 1999 .- - p.
8. Datsenko L.Y., MolodkinV.B., Osinovski M.E., *Dynamical scattering of x-rays by real crystals.* (In Russian) Naukova dumka . 1988, p196.
9. Kato N./Acta Cryst .-1992 .- A48 ,-p.829
10. Fukamashi T. ,Kawamura T. , *Acta cryst* **A49** 38 (1993)
11. Gavrilova E.N. , Kislovski E.N. , Molodkin V.B., Olikchovski S.I., *Metallofizyka* 3- p 70 (1992)
12. Bar`yakhtar V.G., Gavrilova E.N., Molodkin V.B., Olikhovskii S.I., *Metallofizyka.* **14** n3 68 (1992)
13. Pinsker Z.G. *X-ray crystal optic* , M: Nauka, 390 (1982) (in Russian)
14. V.B.Molodkin, V.V.Nemoshkalenko, S.I. Olikhovskii, E.N.Kislovski, O.V.Reshetnyk, T.P. Vladomirova, V.P. Krivitsky , V.F. Machulin, I.V. Prokopenko, G.E.Ice, B.C.Larson, *Theoretical and experimental principles differential –integral triple-crystal x-ray diffractometry of imperfect single crystals* / preprint UNSC (Ukrainian national syncrotron centre) 3 1998, Kyiv 1998.p.23

AUTHOR INDEX

A

Ankudinov, A. L., 105
Arvanitis, D., 140
Aryasetiawan, F., 85

B

Baberschke, K., 140
Benfatto, M., 30
Bianconi, A., 74
Blasco, J., 20

C

Cuozzo, M., 45

D

Dalba, G., 148
Datsenko, L. I., 153
de Groot, F. M. F., 3

E

Ebert, H., 110

F

Fadley, C. S., 123
Fornasini, P., 148
Freeland, J. W., 140

G

Garcia, J., 20
Garcia de Abajo, F. J., 123
Grisenti, R., 148

H

Harder, R., 130
Hlil, E. K., 45
Hodeau, J. L., 20

I

Idzerda, Y. U., 140

J

Joly, Y., 20, 30, 45

K

Kim, C., 97
Kino, H., 85
Klad'ko, V. P., 153
Kotani, A., 13

L

Lograsso, T. A., 140

M

Machulin, V. F., 153
Manninen, S., 153
Mavromaras, A., 110
Minár, J., 110
Miyake, T., 85
Moritz, W., 130

N

Natoli, C. R., 30, 45
Nesvishskii, A., 105

P

Popescu, V., 110
Poulopoulos, P., 140
Proietti, M. G., 20
Prokopopenko, I. V., 153

R

Rehr, J. J., 105
Renevier, H., 20
Rocca, F., 148
Rogalev, A., 140

S

Saini, N. L., 74
Saldin, D. K., 130
Sanchez, M. C., 20
Sandratskii, L., 110

Schattke, W., 57
Schlagel, D. L., 140
Shneerson, V. L., 130
Subias, G., 20

T

Terakura, K., 85

V

Van Hove, M. A., 123
Vogler, H., 130

W

Wende, H., 140
Wilhelm, F., 140